Practical Plant Identificatio

Including a Key to Native and Cul
North Temperate Regions

Practical Plant Identification is an essential guide to identifying flowering plant families (wild or cultivated) in the northern hemisphere. Details of plant structure and terminology accompany practical keys to the identification of 318 of the families into which the flowering plants are currently divided. Specifically designed for practical use, the keys can easily be worked backwards for checking identifications. Containing descriptions of families and listings of the genera within, the book also includes a section on further identification to generic and specific levels.

A successor to the author's *The Identification of Flowering Plant Families*, this new guide is fully revised and updated, and retains the same concise, user-friendly approach. Aimed primarily at students of botany and horticulture, this is a perfect introduction to plant identification for anyone interested in plant taxonomy.

James Cullen is Director of the Stanley Smith (UK) Horticultural Trust. He is also holder of the Royal Horticultural Society's Gold Veitch Memorial Medal, which is awarded to people who have helped in the advancement and improvement of the science and practice of horticulture.

Practical Plant Identification

Including a Key to Native and Cultivated Flowering Plants in North Temperate Regions

JAMES CULLEN D.Sc.
Director, Stanley Smith (UK) Horticultural Trust

CAMBRIDGE UNIVERSITY PRESS
Cambridge, New York, Melbourne, Madrid, Cape Town,
Singapore, São Paulo, Delhi, Tokyo, Mexico City

Cambridge University Press
The Edinburgh Building, Cambridge CB2 8RU, UK

Published in the United States of America by Cambridge University Press, New York

www.cambridge.org
Information on this title: www.cambridge.org/9780521678773

First published 2006

A catalogue record for this publication is available from the British Library

ISBN 978-0-521-86152-6 Hardback
ISBN 978-0-521-67877-3 Paperback

CONTENTS

ILLUSTRATIONS

PREFACE

This current book is a development of ideas put forward in the four published editions of *The Identification of Flowering Plant Families* (1965, 1979, 1989, 1997). The modest success of this book, and its persistence in print for such a long period, has led me to rewrite it completely, taking account of relatively recent developments in the recognition of flowering plant families.

Plant taxonomy is a free and unregulated subject, open to contributions from researchers and students of all kinds; because of this a vast range of opinions co-exist at any one time and to the casual observer the situation may appear chaotic. However, at any time there is a general, unregulated and undefined consensus as to what the taxonomic system should be. This consensus, which is expressed in the use of the various levels of the taxonomic hierarchy in Floras, revisions and other studies involving taxonomic practice, slowly changes in response to new ideas. Over the past 15 years, new ideas of what the family level means have become current: present-day authors are happy to recognise more groups as families than was the case 40 years ago, when the first edition of *The Identification . . .* was published.

The aim of the book is to provide a means of accurate identification of the flowering plant families native in north temperate regions, and to indicate how to proceed to identification of the genera and species; the assumption is made that the user has a general knowledge of plant structure, but is otherwise not particularly expert at identification or taxonomy. The teaching of plant structure and basic classification (and the familiarity with the basic literature) has declined considerably over the past

40 years while the rise of other disciplines in botany has continued, so that there is now a large population of students, scientists and others who may need the means for accurate plant identification, but who haven't the experience to do this without help. This book is intended to provide at least the first stage of the process.

James Cullen

ACKNOWLEDGEMENTS

This book is the product of many years' study and improvement on the basis of advice and information received from many sources. In its earlier format (as *The Identification of Flowering Plant Families*, edns 1–4) I acknowledged the help of very many botanists who had helped with the correction of wrong or misleading facts. These botanists are too numerous to list individually; I hope they will accept this general acknowledgement. In the preparation of all these editions I made extensive use of the collections at the Royal Botanic Garden, Edinburgh, and am grateful to all the staff there, especially Sabina Knees and Suzanne Cubey, who provided both botanical and general assistance. In the preparation of the book in its current format I have made extensive use of the collections at the Cambridge University Botanic Garden; I am equally grateful to all the staff there, especially Professor John Parker, the Director, and to the many garden trainees and students who have attended the Botanic Garden's Summer Taxonomy Course over the past 10 years.

The illustrations used in this book were originally prepared for the second edition under the earlier title by the late Rosemary Smith of RBG Edinburgh.

The original idea for the book came from the late Professor Peter H. Davis of the Department of Botany at Edinburgh University, and he was the senior author for the first three of its earlier editions. My debt to him is enormous.

INTRODUCTION

The family is one of the ranks of the taxonomic system, as developed by plant taxonomists over the past two hundred years, that is important in the accurate identification of plants. In fact, it is the first stage (leaving aside the distinction between dicotyledons and monocotyledons) in the process of using the taxonomic hierarchy for the purposes of of identification. Identification means the finding of the correct name for an unknown plant, not as an end in itself, but as a means of access to all the information so far available about that plant. This is the primary purpose of the taxonomic system; later uses of this system, such as that purporting to show evolutionary relationships among the various taxa (phylogeny), are secondary and frequently not particularly helpful in the achieving of the primary aim.

In the flowering plants (Angiospermae), the number of families (often also called natural orders before 1905) was originally quite small (A. L. de Jussieu's *Genera plantarum* of 1789 contained exactly 100 families, although the precise significance of this number is not known), and the original idea was that all of the families could be known by a single person. Further developments in taxonomy, dependent on the increasing exploration of the world and its plants, showed that this small number was untenable and that more families would have to be recognised. There has been a general slow increase in the number of families since; at present, over 750 families have been proposed to cover the information presented by the flowering plants as a whole. The current book takes something of a middle way between these extremes.

Because the family is a complex level of the hierarchy, synthetic in the sense of grouping together genera that are thought to be similar ('related') to each other, and analytic in the sense that the synthetic groups so recognised must be distinguishable from each other in some ways, a difficulty immediately arises: some authors favour the synthetic approach, others the analytic, resulting in the confusing situation that no two authors of taxonomic systems recognise exactly the same families. This problem has been compounded by the rise, in the past 20 years, of molecular taxonomy (the use of DNA sequences as information for classificatory purposes), which has produced an avalanche of new family and other arrangements which have not yet been properly evaluated or absorbed into the broader taxonomic system.

Any taxonomist with the knowledge and enthusiasm can produce his own system; because there is no regulation covering the acceptance of these systems, many co-exist, some older, some newer. Older systems persist, for instance, in the arrangement of herbaria, in which it is expensive and inconvenient to change physically from one system to another: for example, the major herbaria in Britain (Royal Botanic Gardens, Kew; Royal Botanic Garden, Edinburgh; Natural History Museum) are essentially organised according to a system proposed in the second half of the nineteenth century by George Bentham and Joseph Hooker (see entry 4 in the annotated bibliography, p. 262). In practice, this system has been much modified in detail, but the basic organisation remains. On the European continent, however, most herbaria are arranged according to a system developed by A. Engler around the turn of the nineteenth and twentieth centuries (see entry 11 in the annotated bibliography, p. 263); again there has been much modification in the details, but the main organisation persists.

Similarly, Floras are generally written to follow some taxonomic system current when they were being written. Very rarely are they alphabetical or arranged in some other arbitrary way.

This general looseness, though providing flexibility for the taxonomist (and phylogenist), is not particularly helpful to the general user. It means that, in comparing families from herbarium to herbarium, or book to book, it is necessary to know what system each is following. This often requires careful study of the text and use of indexes and lists of genera for each family.

The present book includes 326 families native or cultivated in north temperate regions (in practice the southern limit of the area is approximately 30° N, thus excluding all of Mexico and Florida in the New World and most of subtropical India and China in the Old World). The system in which they are presented is something of a mongrel; this has been done deliberately to allow for easy comparison with systems old and new. The families recognised, and the genera included in them, are essentially taken from the classification used at the Royal Botanic Gardens, Kew, as published in R. K. Brummitt's *Vascular Plants: Families and Genera* (1992) - entry 6 in the annotated bibliography, p. 263 - which is a complete listing of all the genera in their families as recognised at Kew. The completeness of this volume means that any genus can be found and assigned to its family. In this present book, there are a number of minor divergences from the Brummitt list: these are noted in the text.

The order in which the families are listed, and their numbering, is based on the system proposed by H. Melchior in *Syllabus der Pflanzenfamilien*, edn 12, volume 2, 1964. This system, which is a development of the Engler system mentioned above, was widely influential in the middle years of the twentieth century; many Floras made use of it, including *Flora Europaea* (1964–1980, edn 2, vol. 1, 1995) and *The European Garden Flora* (1984–2000). The families recognised by Melchior are numbered 1–258; families subsequently split off from these families are indicated by a letter after the appropriate number; thus, **229d Aphyllanthaceae** shows that this family was included in family 229 (Liliaceae) in Melchior (and in edn 4 of *The Identification of Flowering Plant Families*). Each family also has a running number.

The 326 families covered by this book are of very varying sizes: a family may consist of a single genus that consists of a single species, e.g. the Scheuchzeriaceae, which contains the single genus *Scheuchzeria*, which contains the single species *S. palustris*. On the other hand large families contain some hundreds of genera and several thousand species (e.g. Orchidaceae, Compositae, Rubiaceae, Euphorbiaceae, etc.). This variation in size means that some families are considerably more variable than others, and that characters that are diagnostic in some cases are not so in others: hence many families key out more than once in the key.

The short descriptions of the families are intended both as a check on identification and as a terse presentation of the important family characters. These descriptions refer to the families as wholes, not just to those representatives covered by the key. The distribution of each family is given, although without great detail. For each family a list of all the genera meeting the criteria for inclusion in this book (see above) is given. This should help the user to understand the limits of the families as recognised here, and to make comparisons with other books. For further details on the various genera, see the book by Brummitt mentioned above.

The long chapter on plant structure (*Examining the plant*, pp. 5–50) provides a brief survey of plant structure and its associated terminology, as used in the key and descriptions. This should be studied carefully by inexperienced users. The short section entitled *Further identification and annotated bibliography* (pp. 260–268) is intended to help the user to proceed further with the identification process.

It remains to be stressed, yet again, that the purpose of accurate identification is the finding of a name, which leads on to all the information available about the plant bearing that name; if the identification is accurate, then the information found will generally be accurate as well.

Examining the plant: a brief survey of plant structure and its associated terminology

The identification of plants is carried out on the basis of the information available about the plant in each particular case. In most situations this information will be derived from a specimen of the plant itself, either whole (if the plant is small, or if it is being examined *in situ* while growing) or a part (generally a stem or twig, with or without flowers or fruits), and consists of the structure displayed by the specimen (its morphology) together with other information that might be available (e.g. where the plant came from originally). On the basis of this information one can make use of the keys in this book to obtain an accurate identification of the family to which the specimen belongs. In order to do this, the specimen has to be observed carefully, so that the structure it displays, and the terminology needed to describe it, are properly understood. The rest of this chapter provides a very brief survey of flowering plant morphology in so far as it is needed for family identification. Each new term is italicised at its first appearance, and appears in the Glossary (p. 269). Further information can be found in textbooks of botany, in Bell, A. D., *Plant Form*, Oxford (1991), which is extremely well illustrated with fine photographs, and in other glossaries, such as Hickey and King, *The Cambridge Illustrated Glossary of Botanical Terms*, No. 19 in the annotated bibliography (p. 264).

The level of detail included here covers what can be seen with the naked eye or with the aid of a hand-lens magnifying 10–15 times, or other directly perceptible characteristics (e.g. scent). In making classifications, plant taxonomists may use not only these characteristics but also others that require more complex equipment: both light and electron microscopes, as well as various pieces of laboratory equipment. The classifications so produced, however, are generally expressible at the simple naked-eye morphological level, even though their information base is much wider than this.

I Duration and habit

Plants may be *herbaceous*, that is, they produce little or no persistent, woody tissue above ground and their stems are soft and without obvious bark, or *woody*, with persistent, hard, aerial twigs, which usually possess obvious bark.

Herbaceous plants may persist for just one growing season: the seed germinates, grows into a plant which produces flowers, fruits and seeds, and then dies off, all within one continuous span of a year or less. Such plants are known as *annuals*. In north temperate areas, most annuals germinate in the spring and die off in the autumn or early winter. A few, such as *Arabidopsis thaliana* (thale cress) or *Capsella bursa-pastoris* (shepherd's purse), both fairly common garden weeds, germinate in autumn, pass the winter as small rosettes of leaves near the ground, and flower in the following spring; such plants are known as *winter-annuals*. Annuals may be recognised by the following features: they have small, slender roots (often surprisingly small for the bulk of the plant above ground), and almost all the branches produce flowers or inflorescences, particularly towards the end of the growing season.

Herbaceous plants that last for two seasons are callen *biennials*; they usually germinate in the spring and produce a rosette of leaves during the first year, which persists through the subsequent winter and then produces flowering shoots, fruits and seeds during the following spring and summer, after which the whole plant dies off. As in annual plants, most of the shoots eventually produce flowers and fruits. The distinction between annuals and biennials is not always clear-cut, especially with plants seen on only one occasion in the wild (in a garden, of course, plants can theoretically be observed through their life-cycles). Biennials can, however, usually be recognised by the co-existence of 1-year-old non-flowering rosettes growing among flowering plants of the same kind.

There are some plants that act like biennials in that they first produce a rosette of leaves, which does not flower immediately; this rosette may persist for several years, 5 or 6 in some species of *Meconopsis* (the Himalayan poppy), 50–100 in some species of *Agave*. Such plants are described as *monocarpic*.

Herbaceous plants that persist for several seasons, flowering every year (except sometimes their first), are called *herbaceous perennials*. Their flowering stems die back to ground level (or near it) every winter, and the plant persists as underground parts, which can become quite woody. Occasionally, in some species of herbaceous perennials, leaf rosettes persist at ground-level through the winter. In all, however, some shoots in each year do not produce flowers and fruits, but form the basis of growth for the subsequent season.

Woody plants have aerial, woody stems and twigs, which persist through several to very many winters. The shoots may be thin and wiry or thick and massive, but whatever their size they bear buds (often protected by waxy or shiny scales), which allow for further growth during favourable seasons, and often have noticeable bark (in plants from areas where growth is possible throughout the year, buds as such are strictly not present, the growing points producing new leaves as and when appropriate).

Subshrubs are generally small, low plants with thin, wiry, woody stems; they can be easily mistaken for herbaceous perennials, but are distinguished by the persistent, woody shoots above ground, as seen, for example, in many species of heather (*Erica*). In Latin, such plants are known as *suffrutices*, and the adjective derived from this, *suffrutescent*, is sometimes used in the botanical literature to describe them.

Shrubs are larger woody plants with obvious, persistent branches. Generally they have several main stems which tend to arise at, or from near, ground level. There is no sharp distinction between shrubs and trees; the latter are generally larger and usually have a distinct trunk or *bole* (sometimes several) which tends to raise the branches well above ground level (although trunks may well branch near the base when the plant is older). These usages of the terms shrub and tree correspond fairly well with the terms as used in common speech, but the degree of precision is somewhat greater.

A few woody plants behave *monocarpically* (see above) in that they build up not a rosette of leaves, but a plant-body that is a tree or shrub; this bears flowers only once, and then the whole plant dies, at least to the level of the underground parts (flowering itself may last for several years). Such behaviour is described as *hapaxanthic* and is seen in some palms and bamboos.

Climbers, which climb by structural means (*tendrils*, etc.), may be either herbaceous or woody. A few (mainly tropical) plants can be shrubs if no support is available, or climbers if it is; such species, if support becomes available during the lifetime of the plant, can begin as a shrub, continue as a climber and finally succeed as a tree.

A small number of plants are *parasitic* on other plants; that is, they draw all or most of their nutrition from the host plant. Such parasites tend to have very reduced plant-bodies, lack chlorophyll and generally have a rather simplified vegetative morphology. They should not be confused with *epiphytes*, plants that grow on other plants without extracting any nourishment from the plants they grow on. Epiphytes tend to have 'normal' (i.e. not reduced) vegetative morphology. A small number of plants are *half-parasites* in that they draw nutrients from host plants but also support themselves to some extent by photosynthesis (e.g. species of *Melampyrum*). Again, a small number of plants are *saprophytes*, absorbing complex chemicals from the soil and its fungal contents rather than making them themselves.

II Underground parts

These are not extensively used in plant identification, because they are not often seen, but their importance in the distinction between the various kinds of herbaceous plant has already been noted. There are several distinct types of underground part.

Roots anchor the plant in the soil and absorb water and minerals. They generally grow downwards and outwards, are never green, and never bear leaves or buds. The first root of the seedling, the primary root, may persist, growing in length and thickness and bearing many branches, forming a taproot system (as in most dicotyledons), or the primary root may not last long, its functions being taken over by roots produced from buds at the bases of the stems (*adventitious* roots), forming a fibrous root-system (as in most monocotyledons). Some plants bear roots that become swollen and act as food-storage organs (e.g. the carrot); such organs are known as *root-tubers*.

Some plants, mainly those that grow epiphytically, produce aerial roots from adventitious buds on the stems. These roots may descend to the soil where they absorb nutrients and water, as in some tropical orchids and the familiar Swiss-cheese plant (*Monstera pertusa*), or they may simply hang in the damp atmosphere and absorb moisture (as in many tropical orchids). In a few cases, aerial roots have other functions (e.g. the climbing roots of ivy, *Hedera helix*).

Underground stems look superficially like roots, but they bear buds and small reduced leaves (*scale-leaves*) and frequently grow horizontally or almost so; the buds may produce branches or may produce upwardly growing shoots. Such underground stems are known as *rhizomes*; they may become swollen for food-storage, when they are known as *tubers* (as in the potato, *Solanum tuberosum*). Rhizomes occasionally extend above ground, looping and then rooting at a point some distance from the base of the parent plant. Such aerial rhizomes are known as *stolons* or *runners*; the strawberry plant (*Fragaria*) provides a familiar example.

Rhizomes that are very short, swollen, bulb-like and upright are known as *corms*, as seen in the species of *Crocus*. *Bulbs* are complex organs made up of modified roots, stems and leaf-bases. Most of the bulb consists of the swollen bases of leaves, which overlap and enfold each other (as in the onion, *Allium cepa*) and are attached at the base to a flat or broadly pyramidal plate, which is the effective stem (bearing roots on its outer side). The outermost leaf-bases tend to be fibrous or papery, and serve as protection for the more delicate tissues within.

III Above-ground parts

These are the most conspicuous parts of the plant and indeed form what is commonly thought of as 'the plant' itself. They are attached to the root, mostly at or near soil level, by a transitional zone that is sometimes called the *stock* or *caudex*. The aerial parts may be very extensive, consisting of various organs, which will be described here serially from the base upwards.

1 Stems

These are the main supporting structures of the plant above ground, bearing the buds, leaves, flowers and fruits. They are generally *terete* (circular in section), although square sections are found in the Labiatae and a few other families, and the stems can be winged or with other outgrowths. They may be erect to horizontal, sometimes erect near the base and then arching over so that the tips are pendulous. With woody plants the term 'stem' is rarely used, the words 'trunk', 'branch', 'twig' or 'shoot' being used depending on the size of the part in question. The term 'stem' is used here in a more precise sense than it is in general English. For instance, the stalk on which a dandelion (*Taraxacum*) flower-head is borne is technically not a stem (it bears neither leaves nor buds), although it is often described so in common speech. Such stalks, which occur in many herbaceous plants, especially those with bulbs, bearing the flowers above a ground-level rosette of leaves, are correctly termed *scapes*.

Most stems bear a bud or growing point at the tip, and produce elongating growth by means of this, bearing leaves, etc., at varying distances from each other, with leafless parts of the stem between; in some plants there is a stem-dimorphism, with some stems like those described above, others, which tend to bear the active leaves, scarcely elongating, forming condensed short-shoots (e.g. in many species of *Berberis*).

2 Leaves

Leaves are present in most plants and form an extremely variable set of organs; they produce many features that are important in identification. In most plants they are borne directly on the stems, twigs or branches, but they may also be borne in basal rosettes or on short-shoots (see above).

The point on the stem, twig or shoot at which a leaf, a pair of leaves, or a whorl of them is borne is known as a *node*; the leafless parts of the shoot, between the nodes, are known as *internodes*.

Duration

Leaves may last for only a single growing season, emerging from buds in the spring and dying off and falling during the autumn; such leaves are *deciduous*. Alternatively, they may last for several seasons, when they are *evergreen*. Deciduous leaves are usually thin and papery or parchment-like in texture, whereas evergreen leaves are usually thicker and somewhat leathery (*coriaceous*) or needle-like (*acicular* or *subulate*). In some species the leaves are half-evergreen, some falling after one season, others persisting for longer.

Attachment to stem

The leaves are attached to the stems at the nodes. There may be a single leaf at each node, a pair (on opposite sides of the stem) or a whorl. When there is a single leaf per node, and the leaves are arranged along a spiral, then they are described as *spirally arranged*. If the leaves are similarly one per node, but successive leaves are on opposite sides of the stem, then the leaves are *alternate* or *distichous* (sometimes no distinction is made between spirally arranged and alternate, and both types are described as alternate, although this is not strictly accurate). If there are two leaves at each node, they are generally arranged on opposite sides of the stem, when the arrangement is termed *opposite*; when successive pairs of leaves are at 90° to each other, the arrangement is described as opposite and *decussate*. When there are several leaves at each node, the arrangement is described as whorled. Care must be taken to distinguish between spirally arranged leaves borne close together and forming *false whorls* (as in many species of *Rhododendron*) and those arranged in true whorls. Leaves may be borne in rosettes at the base of the scapes, as in the dandelion (*Taraxacum*) and many other species. In such rosettes the internodes are very reduced and the leaves are borne close together. It is, however, generally easy to see that the leaves in such rosettes are spirally arranged.

The leaf may be attached to the node by a long or short stalk known as a *petiole*; this may be sharply distinguished from the broader part (the *blade* or *lamina*), or the blade may taper imperceptibly into it. Leaves that have no petioles are described as *sessile*.

The upper angle between the petiole (or leaf-base if the leaf is sessile) and the stem to which it is attached is known as the *axil*. In each axil there is a bud or a branch that has developed from a bud; some plants have multiple buds, one above the other, in each axil (as in some species of *Spiraea*).

Division of the blade

The leaves of most plants are unitary structures, not divided up into separate segments or *leaflets*. Undivided leaves, which may be variously lobed or toothed, are described as *simple* and as *entire* when they have no sign of lobing or toothing (i.e. the margin is smooth and unindented, as in privet, *Ligustrum vulgare*). Toothed leaves have margins that are incised slightly, either regularly or irregularly; when the toothing is sharp and regular, like that of a saw, the leaves are described as *serrate*; when each tooth is itself irregularly toothed, the term *biserrate* can be used. When the teeth are larger and not so regular, the margin is described as *dentate*; dentate leaves can be either regularly or irregularly so. When the teeth are rounded rather than sharp, the margin is described as *crenate*. All the leaves described above have only their margins divided. Lobed leaves are incised to at least one-third of the distance from margin to midrib; the terminology used to describe them is related to that for *compound* leaves (leaves divided to the midrib into distinct and separate leaflets), so these are discussed together below.

Compound leaves are made up of two or more quite separate segments, which are called *leaflets*; leaflets may be stalked, when they are described as *petiolulate* (as in the common house-plant *Schefflera*), or, more commonly, they are without stalks (i.e. sessile). To the superficial observer, leaflets can easily be confused with individual leaves; the essential difference is that at the base of a leaf there is a bud (or a branch that has developed from a pre-existing bud) in the axil, whereas in leaflets neither of these organs is present.

Leaves may be divided into leaflets in two ways: the leaf may be arranged like a feather, with the leaflets arranged parallel to each other along the sides of the main axis (*rachis*), which is a continuation of the petiole; or all the leaves may arise from the same point as the top of the petiole. The first type of division is known as *pinnate*, the second as *palmate*. In pinnate leaves, the leaf may end in a single terminal leaflet (when the leaf is described as *imparipinnate*), or there may be no obvious terminal leaflet, when the leaf is described as *paripinnate*. In some climbing plants the terminal leaflets of the pinnate leaves are replaced by tendrils (see below).

In the case of a plant that has leaves consisting of three leaflets only, such as most clovers (species of *Trifolium*), the leaf could theoretically be either pinnate or palmate; to avoid confusion, such leaves are described as *trifoliolate* (sometimes mis-spelled as 'trifoliate').

In all compound leaves, the leaflets are separate from each other to their bases. Leaves can, however, be simple but lobed; the lobing may extend only as far as one-third of the distance from margin to midrib, or it may extend almost (but not quite) to the midrib. Leaves that are lobed pinnately from one- to two-thirds of the distance from margin to midrib are known as *pinnatifid*; if the lobing reaches further than two-thirds of this distance, but does not quite spearate the blade into distinct leaflets, the leaf is described as *pinnatisect*. Similarly, the terms *palmatifid* and *palmatisect* are used to described leaves that are lobed in a manner reminiscent of palmate division.

The leaflets of compound leaves (or the lobes of lobed leaves) can themselves be toothed, lobed or further divided into leaflets of the second degree (and these, rarely, into leaflets of the third or even fourth degree). Most commonly such division is found in pinnate leaves, which are then known as *bipinnate* (or doubly pinnate, as in the florist's mimosa, which is actually a species of *Acacia*), *tripinnate*, or quadripinnate.

Shape

The shapes of leaves (or, when appropriate, leaflets) are infinitely variable, and a huge terminology has been developed to cope with this variability. This terminology is based on the overall shape considered as the length relative to the breadth, and the position of the broadest part (whether at the middle of the leaf or in the upper or lower thirds). It is not important for identification at the level of the family and so is not discussed further here.

Stipules

Around the point at which the petiole or leaf attaches to the stem there may be two outgrowths known as *stipules*. Their presence or absence is of great importance in family identification. If they are present, they can be extremely variable in form; some are very small and inconspicuous, others large and leaf-like. In a few species (e.g. *Lathyrus nissolia*) they are larger than the rest of the leaf and form the main photosynthetic organs.

Stipules may be separate from the petiole, or joined to it, as in the rose (*Rosa*), when they are described as *adnate* to the petiole (the word adnate is always used to indicate that organs of differing types are joined together, in this case petiole and stipule; the word *connate* is used when organs of the same type are joined together, e.g. connate petals forming a tubular corolla). When the leaves are opposite, the stipules may form a pair on either side of the stem, between the attached bases of the leaves, as in many temperate members of the family Rubiaceae. In some members of this family (e.g. bedstraws, species of *Galium* and its allies) the four stipules belonging to each pair of leaves are as large as the leaves and similar in appearance and structure, so that it appears that the leaves are borne in whorls of six (rarely four by suppression of two stipules); examination of the position of the buds reveals which of the six are genuine leaves.

Stipules may be very quickly deciduous, falling almost as soon as the leaves have expanded. This happens in many tree species (e.g. *Betula, Quercus*) and the bases of the mature leaves must be examined carefully to see the small scars left by the fallen stipules. In *Magnolia* each young leaf-bud is completely wrapped by its stipules, which provide protection against severe weather.

Leaf-scars

The importance of observing the scars left by fallen organs has already been mentioned above. In woody plants, scars are also developed when the leaves fall; the position, shape and form of these scars can be important in identification, particularly in winter when other features are not available.

Veins

Leaves contain a network of harder tissue in the form of veins, which provide mechanical support for the leaf and carry the water- and food-material-bearing tissues (the *xylem* and *phloem*), together with other structures forming the *vascular bundles*. In the leaves of most dicotyledons there is a prominent midrib, which enters the leaf from the petiole and runs up the median line to the tip, giving off secondary branches, which are themselves branched and ultimately form a network. Venation of this type is known as *reticulate*. Another common type of venation is found mainly in monocotyledons and involves several more or less equivalent veins entering the leaf from the petiole or base and running independently to the margins; these *parallel* veins are usually interconnected by smaller veinlets. Reticulate venation is essentially produced in leaves that grow to their final sizes mainly around the margins, whereas parallel veins are produced in leaves that grow mainly at the base. In very thick or fleshy leaves it is usually not possible to see the venation, although holding the leaf up to the light can be helpful.

Ptyxis

This is the overall term used to describe the various ways in which young leaves are compressed to pack in the vegetative buds. It can be examined by sectioning a bud transversely, or by observing the very young leaves as they emerge. This can be a helpful feature in identification in late winter or early spring. The most common type has the leaf with its sides folded upwards along the midrib, with the sides parallel and close together; this condition is described as *conduplicate*. In larger leaves that are lobed or divided, each lobe or leaflet can be folded in this way, producing a pleated effect, for which the term is *plicate* (e.g. *Alchemilla*). Alternatively, the leaves of many species, especially monocotyledons, are rolled up into a tube in the bud, with one margin exterior, the other interior; this condition is known as *supervolute* and is familiar to people who grow the common house-plants *Monstera pertusa* or *Ficus elastica*. In some other plants, e.g. violets (species of *Viola*), the leaves have their margins rolled upwards and inwards in bud, a condition known as *involute*, whereas in species of *Polygonum* the leaf-margins are rolled downwards and under, a condition known as *revolute*. In some plants (e.g. many Amaryllidaceae) the leaves are flat or lightly curved; other plants have various combinations of the characteristics described above. In the insectivorous plants of the family Droseraceae the leaves are elongate and rolled from the tip to the base, with the upper surface either inside the spiral or outside; in both these cases the ptyxis is described as *circinate*.

There are various other terms used to describe more minor characteristics of leaves; these are treated below under 'Miscellaneous features' (see p. 49).

3 Flowers

The flowers and the fruits that they lead to are the most important parts of the plant from the point of view of identification. They are also, of course, the most important parts from the point of view of the attractiveness of the plant. However, this is all problematic, because for long periods of the year any individual plant will have neither flowers nor fruits; or, if it has flowers, it is possible that the fruits have not yet developed; or, again, if it has fruits it is probable that the flowers are already over. For the really accurate identification of many plants, both flowers and fruits are needed; both may be available at the later stages of flowering, which is therefore the best time to attempt identification. If the plant is growing in a garden, it can be examined twice, once in flower and once in fruit. In some cases (e.g. with many species of *Cotoneaster*) it is necessary to press flowering specimens so that these can be available when the plants have mature fruit, which is generally much later in the year.

A major problem consists in the definition of what a flower is. The word, as used in normal speech, is imprecise: the 'flower' of a dandelion is an inflorescence (a collection of flowers forming a coherent whole). Flowers vary greatly from species to species and it is difficult to find a definition that covers all the cases that occur. Possibly the best that can be done is to say that a flower is usually borne at the top of a long or short stalk (the *pedicel*) or, if stalklesss, has its insertion on some other organ and contains either one or more female sexual organs (*carpels*, see below) or one or more male sexual organs (*stamens*) or one or more carpels together with one or more stamens; there may be other parts associated with these as well, which may be protective in function (generally the *calyx*, composed of *sepals*, which are free from each other or united into a tube at the base), or pollinator-attractive (the *corolla*, composed of individual *petals*, which may be free from each other or united into a tube at the base). Nectar-secreting organs may also be present, represented by one or several *nectaries*, together with organs that are apparently reduced leaves (*bracts* or *bracteoles*, see below). Any of these organ-groups may be absent, depending on the particular type of flower. For instance, the female flower of a spurge (*Euphorbia*) consists of three united carpels (the *ovary*) only, whereas the flower of a catchfly (*Silene*) has a bract, bracteoles, sepals, petals, stamens, a nectary and an ovary.

The number of possible combinations is very great, and the simplest way of discovering what is a flower in any particular plant is to find an ovary (or, if an ovary is entirely absent because the flowers are unisexual and male, a group of stamens which are obviously associated with each other), and to look at the organs immediately surrounding that. All these organ-groups, except for the bracts and bracteoles, tend to occur in whorls of two or more or, in rare cases, in compressed spirals. Even expert botanists are sometimes confused as to what exactly is a flower when confronted with a plant they have never seen before.

Inflorescences

Flowers are usually grouped together towards the ends of the branches of a plant into units that are called *inflorescences*. Again, there is a problem of definition, because inflorescences are extremely variable. In general terms, an inflorescence may be defined as the arrangement of all the flowers on a branch; it is often easier to see an inflorescence than to define exactly what it is. In most plants, as mentioned above, the inflorescences develop at the ends of young branches, but in some, mostly tropical plants, such as cocoa (*Theobroma*) or the Judas tree (*Cercis*), the flowers are borne on the older, woody branches; this phenomenon is known as *cauliflory*). An inflorescence may consist of only a single flower (as in most tulips), or may contain thousands of flowers.

Within an inflorescence, each flower usually has its own long or short stalk (the pedicel, see above) which attaches it to the axis of the inflorescence (known as the *rachis*). The stalk that bears the whole inflorescence (from above the uppermost leaf) is known as the *peduncle*. In plants with a single, *solitary* flower in the inflorescence, there is, of course, no easy distinction between pedicel and peduncle. There are many herbaceous plants in which the leaves are all borne in a rosette at ground level and the inflorescence is borne on a long or short stalk above them (a scape, see above).

The pedicel of a flower is often borne in the axil of a small, leaf-like organ known as a *bract*; in some cases (e.g. *Veronica*) there is little or no distinction between the upper foliage leaves and the lowermost bracts. On the pedicel there may be one, two or even more organs looking like bracts, but usually even smaller:

these are known as *bracteoles* and may bear further pedicels in their axils. Some plants (e.g. heathers, *Erica*) have both bracts and bracteoles, whereas others have bracts but no bracteoles (e.g. *Acanthus*) and yet others have neither (e.g. stocks, *Matthiola*).

The easiest inflorescence to understand is the solitary flower; this is seen, for example, in florist's tulips, where each stem terminates in a single flower. Similarly, scapes may also bear solitary flowers. Although the solitary flower is easy to understand, it is thought to have arisen, during the course of evolution, from various different types of inflorescence.

Aside from the solitary flower, there are two main kinds of inflorescence. In the first kind, the axis of the inflorescence continues to grow for some time, producing flowers as it grows, so that the oldest flowers are at the base of the inflorescence if it is elongate, or towards the outside if the inflorescence is condensed. This type of inflorescence is a *raceme* (adjective *racemose*). Again, there are basically two kinds of racemose inflorescence: if the individual flowers are stalked, the inflorescence is a simple raceme, whereas if they are stalkless (*sessile*) the inflorescence is a *spike*.

In the other main type of inflorescence, the apex of the stem ceases to grow and becomes a terminal flower. Subsequent flowers arise on side-branches (generally borne in the axils of bracts), which originate below the oldest flower (the terminal one). Hence, the oldest flowers are found towards the apex of the inflorescence if it is elongate, or towards the centre if it is condensed. Such inflorescences are known as *cymes* (adjective *cymose*). There are many different kinds of cyme, depending on which individual flowers are suppressed.

Some plants bear compound inflorescences, either a raceme of racemes, a raceme of cymes or a cyme of cymes. Such inflorescences are generally known as *panicles*.

In some plants, particularly those belonging to the large daisy family (Compositae), the individual flowers are quite small and are aggregated together into inflorescences, which can look like individual flowers. These inflorescences are known as *capitula* (or *heads*) and are basically compressed racemes, as is shown by the fact that the outermost flowers mature earlier than the innermost. The bracts of the flowers form a calyx-like structure beneath the flowers, and this is known as an *involucre*. The individual flowers in each capitulum are generally stalkless, attached directly to the flattened or conical apex of the peduncle (the *receptacle*); they may be all of the same form, or the marginal flowers may differ from the central ones.

A further kind of complex inflorescence is the *catkin*, as found in many trees (*Alnus, Betula, Quercus*, etc.). These are complicated structures usually involving racemosely arranged bracts and bracteoles (which may be variously united to each other) and unisexual flowers without corollas. Catkins may be upright or pendulous, and frequently fall as wholes, rather than breaking up and falling as individual flowers.

In a small number of cases, the true nature of the inflorescence is difficult to ascertain, even with detailed morphological studies. Such inflorescences are referred to as *clusters* or *fascicles*.

In considering the individual flowers, it is best to deal with the various organ-groups separately. They are discussed here from the centre of the flower outwards, as this provides the most sensible framework for explanation. When identifying a plant, of course, the flower is generally considered from the outside inwards, the various parts being recognised and counted before the flower is further dissected or sectioned (see below, p. 24).

The ovary

The ovary is present in all flowers except for those which are unisexual and male. It is found at the centre of the flower and forms the uppermost (latest) part in terms of its origin (that is, it is the last part of the flower to be fully formed, and therefore is found at the morphological apex of the flower; this does not mean that it is necessarily the uppermost part of the flower when viewed, as later developments can take other parts of the flower physically above it).

The ovary consists of, or is built up of, individual units known as *carpels*. There may be just a single carpel forming the ovary, or there may be two to many; if there are more than one, then they may be free from each other, or joined together (united, connate) into a compound structure. If the carpels are free from each other, the ovary is described as *apocarpous*; if the carpels are united, then as *syncarpous*. The term *gynaecium* or *gynoecium* is sometimes used as an alternative to 'ovary'.

The simplest case to consider (not necessarily the simplest case from the evolutionary point of view) is, of course, the ovary formed from a single carpel: the *unicarpellate* ovary. The most familiar example is that of the pea pod, as bought in a greengrocer's shop. This is the ovary of the pea flower as it develops into a fruit. It consists of an elongate bag, attached to the flower (and still bearing the remains of the calyx at its base) at one end, and tapering into the narrow style-base (see below) at the other. The bag is the main body of the single carpel, and it contains the *ovules* which later, when fertilised and fully developed, become the *seeds* (the peas). The ovules/seeds hang down on short stalks (*funicles*) from the upper angle of the bag, and are in two rows, although this can only be easily seen when the ovary is young. The end of the pod, when it was younger, tapered to a thread-like portion, tipped by a velvety apex. The thread-like part is known as the *style* and the velvety apex is known as the *stigma*. In order for fertilisation of the ovules to take place, causing them to develop into seeds, pollen of the correct type has to find its way to the stigma, where it begins to grow, as a fine tube, down through the style to reach the ovules. In peas bought from the greengrocer, the styles and stigmas have fulfilled their function and have been shed.

In hellebores (species of *Helleborus*) and species of winter-aconite (species and hybrids of *Eranthis*), and in many other genera, the ovary is made up of several to many carpels, each of which is similar to that of the pea. The carpels are packed closely in a whorl, but they are not at all joined to each other. In buttercups (species of *Ranunculus*) and many other genera, there are similarly many free carpels, but each one is much shorter, and contains a single ovule only (developing into a single seed). Such ovaries are described as *pluricarpellate* as well as apocarpous; they may have two to many carpels and each carpel may contain one to many ovules. Each carpel in ovaries of this type has its own style and stigma; in some cases, the style is reduced so that the stigma is borne directly on the body of the carpel (i.e. the stigma is sessile).

Unicarpellate or apocarpous ovaries are not all that common in the flowering plants; most have pluricarpellate ovaries in which the carpels are joined together into a compound structure in which the boundaries of the individual carpels are blurred or merged. Such compound ovaries are described as pluricarpellate and *syncarpous*. Syncarpous ovaries show varying degrees of joining of the carpels; the carpels themselves may be completely joined but the styles may be free from each other; or the carpels and styles may be joined to each other but the stigmas free; or all three may be fully joined; rarely, the carpels themselves may be almost entirely separate but the styles, at least, are joined together. All of these types, including the last, which is found, for example in the butterflyweed (*Asclepias*), are syncarpous.

In joining together, the individual carpels may retain their individual internal spaces, or these internal spaces may merge into a single internal space. Thus a syncarpous ovary may contain several ovule-containing spaces separated by cross-walls, or there may be a single internal space which contains ovules contributed by various carpels. The ovule-containing spaces are known as *loculi* (singular *loculus*) or *cells*. An ovary containing several loculi is known as *multicellular* or *multilocular* (*2-, 3-, 4-, or more-locular*), whereas an ovary containing only a single space is known as *unilocular* or *1-celled*. An ovary may be pluricarpellate and unilocular and contain only a single ovule/seed, contributed by a single carpel, the other(s) being sterile.

It is important for accurate identification to determine the number of carpels that make up a pluricarpellate ovary. The best way of deciding this is to count the number of stigmas, if they are free from each other; this is generally the same as the number of carpels. If the stigmas are completely united into a unitary structure, then counting the number of cells in the ovary (if they are more than one) will usually provide the same information (in a few families, notably Linaceae, Boraginaceae and Labiatae, the loculi become further divided by ingrowing secondary walls, so that the number of apparent loculi is twice the number of carpels). If the stigmas are united and the ovary is unilocular (as in *Primula*) then it is very difficult to decide how many carpels there are (microscopic study may be required). In such cases, however, experience shows that the number of carpels is usually either two, or the same as the number of petals or corolla-lobes.

In order to see how many loculi the ovary contains, it is usually necessary to section it horizontally across the middle. Such a section will also generally show how many ovules there are in the ovary or in each loculus, and, very often, the way they are attached to the ovary: the *placentation*, which is a very important characteristic (see below). With some plants it may be necessary to cut another ovary longitudinally (along a diameter) to see precisely how the ovules are attached.

The cutting of the ovary, as described above, is an art that has to be mastered with practice, as the ovaries can be very small indeed. In general terms, it is relatively easy to cut a transverse section by using a single-edged razor blade. The section should usually be taken at the widest part of the ovary, and both the cut ends should be looked at. If necessary, further thin sections should be cut from the cut ends, so that the structure can be more clearly seen. Longitudinal sections can be difficult and troublesome. The flower (or just the ovary if it is large enough) should be held between finger and thumb, with the flower in the same position as it was on the plant, with the stem end pointing outwards. A cut should then be made carefully through this, moving the blade downwards as the cut goes through the tissues. In this way, clean surfaces will be left and the ovary will not be squeezed or squashed, as will happen if a sawing motion of the blade is used. In looking at longitudinal sections it should be remembered that the ideal section cuts through the vertical plane of the flower as it existed on the plant; actual sections may diverge from this ideal to some extent and one needs to think clearly about what the section actually shows.

It is sometimes helpful to rub over the ends of the cut sections with the tip of a black, broad felt-tipped pen. The water-based ink is absorbed by the cut surfaces, which therefore stain black, but is repelled by the waxy surfaces of the ovules, which are pale green and stand out from the rest of the stained cut.

Placentation

This is a very important character in the identification of the families, and should ideally be observed in both transverse and longitudinal sections of the ovary (made as above). Although there is some overlap between the different types of placentation, it can be conveniently dealt with under the terms used in the key.

(i) Marginal. This term is used only in those cases in which the carpels are free or the ovary is made up of a single carpel, and describes the condition in which the carpel bears several ovules on its upper suture (e.g. *Caltha, Pisum*, etc.) (see fig. 1*a,b*). If there is only a single ovule in the carpel, it may be borne at the base, when it is described as [*marginal-*]*basal*, or at the apex, when described as [*marginal-*]*apical* (see fig. *3h,k*).

(ii) Axile. In this condition the ovary is made up of two or more united carpels and contains cross-walls (*septa*) which form the

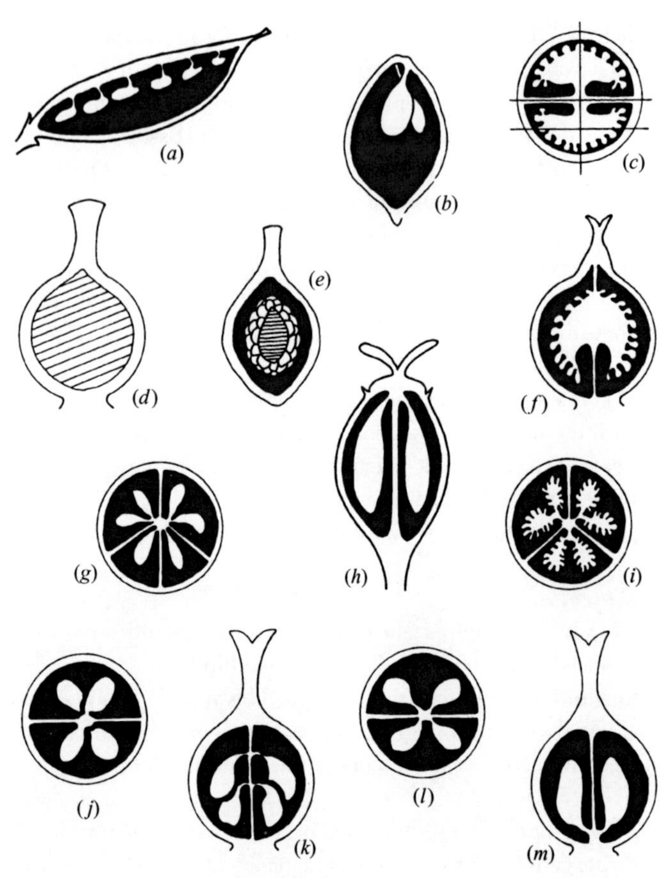

Figure 1. Placentation (see pp. 25–31): (*a*), (*b*) marginal; (*c*)–(*m*), axile. (*c*)–(*f*) Ovules on swollen placentas ((*c*) transverse, (*d*)–(*f*) longitudinal sections; planes of the longitudinal sections indicated in (*c*)). (*g*) Ovules borne on the axis; (*h*) ovules pendulous; (*i*) ovules on intrusive placentas; (*j*), (*k*) ovules superposed; (*l*), (*m*) ovules side-by-side.

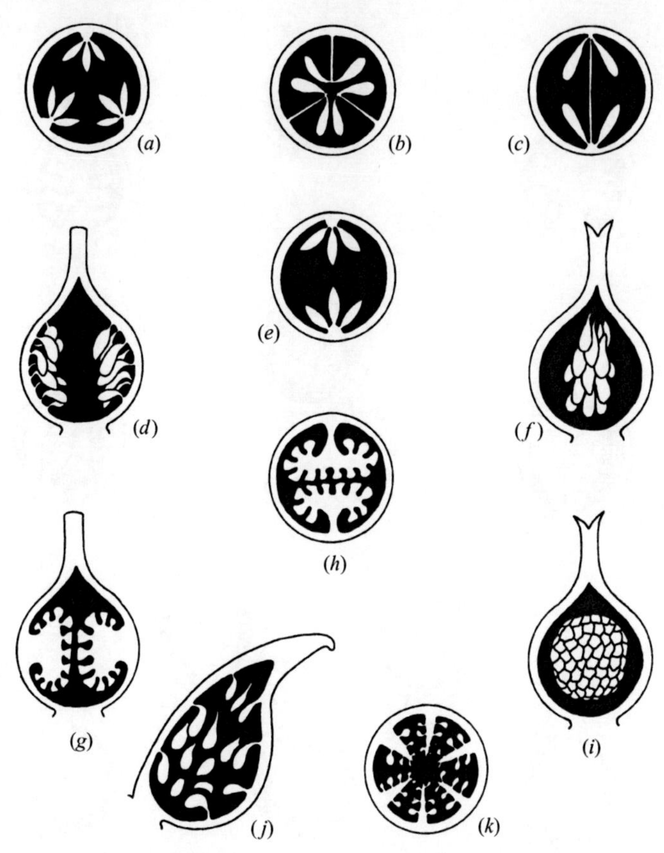

Figure 2. Placentation (see pp. 25–31). Parietal types. (*a*) Ovules on the carpel walls; (*b*) ovules on intrusive placentas; (*c*) ovules on the carpel walls, septum present. (*d*)–(*f*) Ovules on carpel walls: (*d*) longitudinal section through placentas; (*e*) transverse section; (*f*) longitudinal section at right angles to placentas. (*g*)–(*i*) Ovules on intrusive placentas that almost meet in the centre of the ovary: (*g*) longitudinal section through the placentas; (*h*) transverse section; (*i*) longitudinal section between the placentas. (*j*), (*k*) Diffuse parietal placentation.

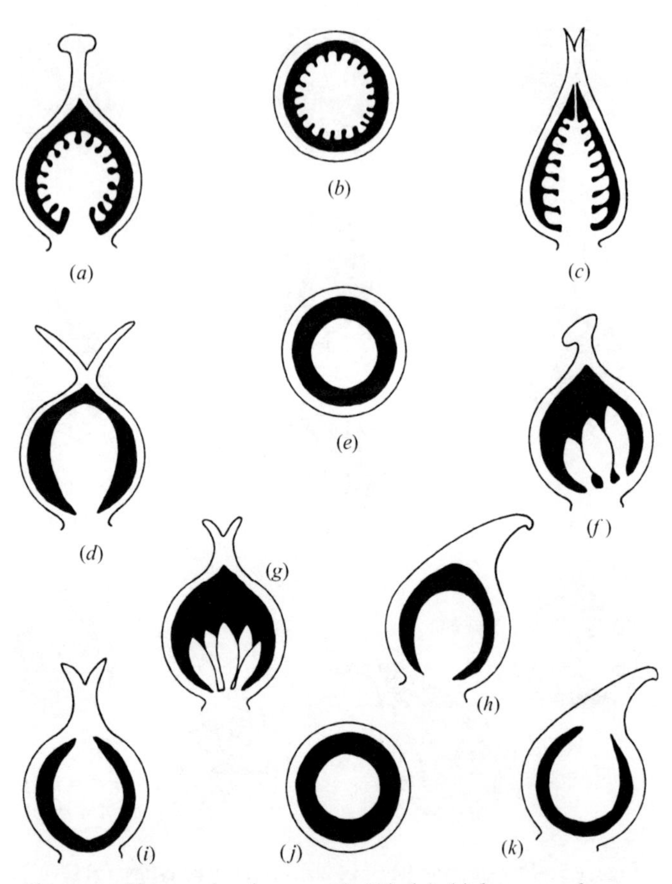

Figure 3. Placentation (see pp. 25–31): (*a*)–(*c*) free-central; (*d*)–(*h*) basal; (*i*)–(*k*) apical. (*a*), (*b*) Ovules free-central; (*c*) ovules free-central, showing attachment of placenta to top of ovary; (*d*), (*e*) one basal ovule; (*f*) ovules on an oblique placental cushion; (*g*) several basal ovules (ovary of united carpels); (*h*) one basal ovule (ovary of free carpels); (*i*), (*j*) ovule apical (ovary of united carpels); (*k*) ovule apical (ovary of free carpels).

loculi or cells. The ovules are borne on the central axis, where the cross-walls meet (e.g. *Narcissus*) (see fig. 1*g*), on swollen placentas (e.g. *Solanum*) (see fig. 1*c–f*) or on intrusive placental outgrowths (e.g. some species of *Begonia*) (see fig. 1*i*). In some families the ovules are reduced to one or two in each cell and ascend from the base (e.g. *Ipomoea*) (see fig. 1*l,m*) or are pendulous from near the apex (e.g. many Umbelliferae) (see fig. 1*b*). Ovules in axile ovaries sometimes occur side by side (*collateral* ovules) as in *Heliotropium* (see fig. 1*l,m*), or one above the other (*superposed* ovules) as in most Acanthaceae (see fig. 1*j,k*). Occasionally axile ovaries are further divided by secondary septa, which grow inwards from the carpel wall as the ovary matures (e.g. *Linum, Salvia*), so that the ovary comes to have twice as many cells as carpels.

(iii) Parietal. This term is used when the ovules are borne on the walls of the ovary, or on outgrowths from them. Several situations may be distinguished.

In the majority of cases, parietal placentation occurs in 1-celled (*unilocular*) ovaries made up of several united carpels, the ovules being restricted to placental regions on the walls, as in *Viola* (see fig. 2*a*), *Gentiana* (see fig. 2*d–f*) or *Ribes*, or on intrusive placenta-bearing outgrowths from them (e.g. *Cistus, Heuchera*) (see fig. 2*b*). Intrusive parietal placentas may almost meet in the middle of the ovary, so the distinction between parietal and axile placentation is not always clear-cut (e.g. *Escallonia, Cucumis*) (see fig. 2*g–i*). Gently squeezing the cut ovary will reveal whether the placentas are united in the centre of the ovary (when the placentation is axile) or whether they just meet there without fusing (when the placentation is parietal).

In a few cases the ovules are borne on the walls of a 2- or more-celled ovary; this is found particularly in the Aizoaceae. In most Cruciferae the older ovary and fruit are 2-celled, but this is because of the development during ripening of a false septum (*replum*) across the ovary (see fig. 2*c*).

Occasionally the ovules are scattered over most of the inner surfaces of the carpels. This situation is distinguished as *diffuse parietal* placentation; it can occur in ovaries with free carpels (e.g. Butomaceae) (see fig. 2*j*) or of united carpels (e.g. Hydrocharitaceae) (see fig. 2*k*).

As shown in the accompanying diagrams (figs. 1, 2 and 3), the side view of axile and parietal placentation can vary greatly according to the vertical plane in which the section has been cut. The longitudinal section should always be considered in relation to a transverse section.

(iv) Free-central. In this condition, the ovules (usually several to many) are borne on a central spherical or columnar structure that rises from the base of the 1-celled ovary made up of several united carpels (e.g. *Pinguicula*) (see fig. 3*a–c*). In most cases a thread of tissue attaches this placental column to the top of the ovary; sometimes this thread is rather stout (e.g. *Lysimachia*) (see fig. 3*c*). Occasionally the ovary may be somewhat septate near the base (e.g. *Silene*) although in most ovaries with free-central placentation such septa break down as the ovary matures.

(v) Basal. Here the ovules (one or more) arise from the base of a 1-celled ovary (or rarely from the base of a solitary or free carpel, see above), as in *Polygonum, Tamarix, Armeria*, etc. (see fig. 3*d,e,g*) or are borne on a basal placental cushion (oblique in *Berberis*) (see fig. 3*f*).

(vi) Apical. In this case the ovule (generally solitary) is attached to the apex of the single cell (or free or solitary carpel, see above), as in *Scabiosa* (see fig. 3*i,j*) or *Anemone* (see fig. 3*k*).

Although it would be possible to describe the ovules in the cells of many septate ovaries as 'apical' or 'basal', the terms have not been used in this sense here. Instead, to avoid confusion, 'pendulous' or 'ascending' have been used, and are considered to be special cases of axile placentation.

The stamens

The male parts of the flower are called *stamens*, collectively forming the *androecium*. They are generally found in one or more whorls outside the ovary in bisexual flowers, but they may be apparently central in the purely male flowers of some unisexual species. Each stamen is a relatively simple structure, consisting of a stalk, the *filament*, which is usually thread-like but can be quite thick, bearing at the top the *anthers* (the pollen-containing parts). These are usually broader than the filament and borne at its apex, but can be narrow and somewhat sunk in the broad filament. The anther is usually made up of four elongated sacs (which may become confluent into two as the anther matures) which contain the pollen. The pollen is usually dry and granular, but in some cases it is sticky (e.g. species of *Rhododendron*) and sometimes aggregated into masses, known as *pollinia*, which are dispersed as wholes (found in orchids, Asclepiadaceae, *Mimosa*, etc.). In a few groups (e.g. most Ericaceae, Droseraceae, Juncaceae, etc.) the pollen grains themselves are not separate but are aggregated into groups of four (*tetrads*); this is difficult to see without the aid of a compound microscope, as are the surface features of the pollen grains themselves, which can be important for identification. These features are, however, beyond the scope of the present book.

The number of stamens, as well as their structure, is important in identification. There may be only a single stamen per flower (e.g. *Euphorbia*), or there may be two to many. In some plants the stamens in each flower are joined together (united, *connate*) sometimes just by their filaments (as in many plants related to the mallows, *Malva*), or by their anthers (as in the family Compositae), or by both; when joined by their filaments, the tube so formed generally surrounds the ovary and style(s).

In order for the pollen to be released, the anthers must open. The most frequent method is for slits to develop down the length of the pollen-sacs; there are generally two such slits, one on either side. In a few groups the anthers open by distinct pores, either really at the top (Polygalaceae, etc.) or apparently so (Ericaceae: during development of these stamens the anther becomes inverted, so that the pores which are apparently at the top are, in fact, at the morphological base). In a small number of cases (*Berberis, Hamamelis* and others) the anthers open by flaps or valves, which may curve upwards and outwards. The opening of the anthers is usually referred to as *dehiscence*; the normal method is known as *longitudinal* dehiscence, that by pores as *porose* or *poricidal*, and that by valves as *valvular*. Porose dehiscence is found in the following families: Ochnaceae, Elaeocarpaceae, Tremandraceae, Polygalaceae, Melastomataceae, Actinidiaceae (part), Ericaceae (most), Pyrolaceae, Monotropaceae, and Mayacaceae, but occurs sporadically in other groups (e.g. in the snowdrop, *Galanthus*, in the Amaryllidaceae, and *Cassia* in the Fabaceae).

In some functionally female flowers, and also in some bisexual flowers, sterile, stamen-like structures may be found either where the genuine stamens might be expected, mixed with the genuine stamens, or outside of them. These organs are known as *staminodes*. In bisexual flowers the origin of the staminodes from the stamens is generally fairly obvious (e.g. *Parnassia, Sparrmannia*), although occasionally staminodes may be difficult to distinguish from petals (e.g. in many *Aizoaceae*).

The perianth

The ovary and stamens form the most important parts of the flower, and indeed many flowers consist of these organs only, either singly or together. However, in most flowers there are other protective and pollinator-attractive structures to the outside of the stamens; these are collectively known as the *perianth* and may consist of distinct calyx and corolla, or of a single series of organs, or more rarely of several series of organs. The organs of the perianth are generally arranged in alternating whorls, although in a few families (e.g. Magnoliaceae) they are arranged in compressed spirals. A perianth with a single whorl of organs is known as *uniseriate*, one with two whorls as *biseriate*, and one with three or more whorls as *multiseriate*.

The case most frequently seen is the biseriate perianth, consisting of two whorls of organs. Those of the outer whorl are known collectively as the *calyx*, whose individual organs are known as *sepals* (when they are completely free from each other) or *calyx-lobes* (when the individual organs are united at the base. A calyx with free sepals is known as *polysepalous*, one with sepals joined at the base as *gamosepalous*; similarly, for the organs of the inner whorl (*petals)*, if the petals are united into a cup or tube at the base they are more correctly referred to as *corolla-lobes* and the corolla as *gamopetalous* (as opposed to *polypetalous*, when the petals are free from each other). It is sometimes difficult to decide whether or not the individual organs of the whorl are joined to each other (*connate*) or not. Generally, a corolla of united lobes falls as a whole, whereas one with free petals falls as separate units.

In general terms the sepals are usually greenish and somewhat leaf-like, at least in texture, and serve to protect the more delicate organs of the flower in bud. Usually, each sepal (or calyx-lobe) has three veins entering it from the base; this is often very difficult to see, either because of the thickness of the sepal itself or because the two outer veins are often much less pronounced than the central, so that the sepals can appear to be 1-veined. The petals, on the other hand, are usually thinner, larger and more brightly coloured than the sepals, usually serving to attract pollinators to the flower. The petals are generally 1-veined from the base, and this can usually be relatively easily seen.

Flowers in which the perianth consists of a single whorl are generally considered to have no petals, the perianth consisting of organs interpreted as sepals. A more neutral terminology for such cases is to describe the organs as perianth-segments or *tepals*. This is the case even if these organs are brightly coloured and petal-like (as in many species of *Clematis*). However, flowers that appear superficially 1-whorled may have two whorls, the calyx being extremely small and reduced (as in many species of *Rhododendron*); this must be looked for very carefully. Some further guidance on deciding whether sepals and petals or perianth-segments are present is given below under '*Horizontal disposition of parts of the flower*' (p. 35).

In the flowers of some species there is no perianth at all. Such flowers are usually wind-pollinated, and the sexual parts are often associated with bracts or bracteoles (see above); this occurs in plants that have flowers in catkins (e.g. birch, oak, willow), or in specialised *spikelets*, such as grasses and sedges.

The symmetry of the perianth is often an important character in identification, and it is one that often causes trouble in interpretation, because nature is never totally symmetrical. Many flowers are built on a radial plan, so that the perianth forms more or less a circle when viewed in outline. Such perianths have many planes of symmetry (sectioning the flower along any diameter will produce two mirror-image halves) and are described as *actinomorphic* or *radially symmetric*. Many common flowers such as the rose and the buttercup have flowers of this type. Others have a perianth with only a single plane of symmetry; such flowers appear to be 2-sided when viewed in outline and are described as *zygomorphic* or *bilaterally symmetric*. Usually the single plane of symmetry is the vertical, but occasionally it is horizontal (as in *Corydalis*). In a few plants, such as species of *Maranta*, the perianth is asymmetric, but the flowers are borne in pairs so that the perianths of the two flowers together are zygomorphic.

As well as the perianth, stamens and ovary can be zygomorphically arranged. This generally occurs when the perianth is also zygomorphic, although it occasionally occurs when the perianth is actinomorphic. In many Ericaceae, for instance, the ten stamens are deflexed downwards in a group, arching towards the lower part of the flower and then arching upwards again towards the corolla opening (stamens and styles *declinate*). In *Gloriosa*, a tropical climber in the Liliaceae, the style is borne at right angles to the axis of the flower.

Nectaries

In some flowers (e.g. *Magnolia, Papaver*) no nectar is produced, but it is found in most. It may be sticky and in small quantity or watery and copious, and is secreted by zones or rings of tissue known as *nectaries*. These may be found on the perianth (on the inner surfaces of the sepals, as in many Malvaceae, or on the petals as in many Ranunculaceae), or on the floral receptacle, either between the petals and stamens or the stamens and ovary, or both, or on the ovary itself. Nectar produced by the nectaries is sometimes held in *spurs* (generally backwardly projecting narrow sacs) developed on one or more of the sepals or petals.

Horizontal disposition of parts of the flower

Flowers are immensely variable in the ways that the parts are arranged; in this section the horizontal arrangement of the parts is described. The vertical arrangement is treated in the next section.

It is simplest to consider first radially symmetric flowers, whose outline forms more or less a circle, in which the stamens are twice as many, or as many, as the sepals and petals. As an example, a flower with parts in fives will be described, but the principles apply to flowers with different numbers of parts. In such a flower, the five sepals will be arranged symmetrically, their apices forming part of a circle, and imaginary lines down their centres will lie at approximately 72° to each other. The five petals will be disposed in the same way, but alternating with the sepals, so that the imaginary line down the centre of each petal lies exactly in between those of two of the sepals. If there are five stamens, they will usually be found on the same radii as the sepals; in a few cases (e.g. *Primula*) this is not so, and the stamens are found on the same radii as the petals, when they are described as *antepetalous*. If there are ten stamens, it is usual for them to be arranged in two whorls of five, those of the outer whorl on the same radii as the sepals, those of the inner on the same radii as the petals. In the rare case when the outer five stamens are on the same radii as the petals and the inner on the same radii as the sepals, the flower is decribed as *obdiplostemonous*; this is very unusual, but occurs in species of *Geranium*. The general principle is that the organs of each whorl (whether they are free or fused) lie on alternating radii; it is thought that suppression of various whorls leads to the situations found in nature. Thus, if a flower has a uniseriate perianth of five segments, with five stamens on the same radii as these, then it is likely that the perianth-segments are in fact sepals (petals having been suppressed or lost during evolution). Similarly, if the perianth has five segments and there are five stamens on radii alternating with those of the perianth-segments, then it is possible either that the segments are petals (sepals having been suppressed or lost) or that they are sepals and one whorl of stamens has been suppressed or lost.

This principle of alternation of radii continues with respect to the carpels in the ovary. When the number of carpels is the same as that of the stamens and perianth-segments it is generally easy to see this, but it may become obscured when the number of carpels is smaller than that of the stamens and perianth-segments (as is often the case).

The same principle can be applied to zygomorphic flowers, although with these the shape of the perianth sometimes causes difficulties in determining the various radii. When this is the case, it is necessary to look at the base of each whorl of organs to see how they are arranged with respect to the others.

In bud, the sepals and petals or perianth-segments are often arranged in characteristic ways. This phenomenon is known as

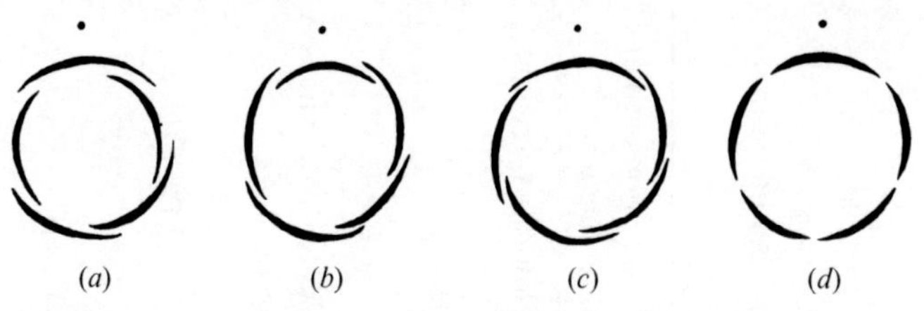

Figure 4. Aestivation types. (*a*), (*b*) Overlapping (imbricate); the details of the manner in which the organs overlap each other is variable and two of the common types are shown. (*c*) Contorted (each overlapping one other and overlapped by one other: a special case of imbrication). (*d*) Edge-to-edge (valvate).

aestivation and is sometimes of importance in identification. In the simplest case, the organs lie edge-to-edge in bud, without overlapping (indeed, sometimes with quite considerable spaces between them). This is known as *valvate* aestivation (see fig. 4*d*). In other cases the various segments overlap each other; the general term for this is *imbricate* aestivation. There are various kinds of imbricate aestivation, depending on the alignment of the various organs. The only one necessary to distinguish for present purposes is that in which each segment overlaps, and is overlapped by, one other segment. Such aestivation is known as *contorted* (fig. 4*c*) and can be seen, for example, in the periwinkle (species of *Vinca*).

Vertical disposition of parts of the flower

Two sets of terms are used in describing the relative vertical positions of the attachments of the floral organs. One (superior/inferior) is used with reference to the position of the ovary with respct to the other floral organs. The other (hypogynous/ perigynous/epigynous) refers to the positions of the other floral organs with respect to the ovary, but is best explained in terms of the apparent fusion of the organs of different floral whorls.

Table 1 *Relationships of floral parts (see fig. 5)*

The 'ring' or 'collar' of tissue mentioned in the table, probably best referred to as a *perigynous zone* if the ovary is superior and as an *epigynous zone* if the ovary is inferior, is often referred to in other literature as 'calyx', 'floral cup', 'floral tube' or 'hypanthium'.

Ovary (G) position	Fig. 5	Insertion of perianth (P or K & C) and androecium (A)	Description adopted here	Description used in older literature
	(*a*)	PA or KCA inserted independently on receptacle (e.g. *Ranunculus*)	PA or KCA hypogynous	Flower hypogynous
	(*b*)	K & C apparently fused at base, A inserted independently on receptacle (e.g. *Tropaeolum*)	K & C perigynous borne on a perigynous zone, A	Various
superior	(*c*)	C & A apparently fused at base, K inserted independently on receptacle (e.g. *Primula*)	K hypogynous, C & A perigynous borne on a perigynous zone	Flower hypogynous A epipetalous
	(*d*)	K, C & A inserted on a ring or collar of tissue which is inserted on receptacle (e.g. *Prunus*)	K, C & A perigynous, borne on a perigynous zone	Flower perigynous
	(*e*)	P & A apparently fused, C absent (e.g. *Daphne*)	P & A perigynous	Various

half-inferior	(*f*)	P & A or K, C & A inserted independently on walls of ovary (e.g. *Paliurus*, some species of *Saxifraga*)	P & A or K, C & A partly epigynous	Various
	(*g*)	P & A or K, C & A inserted independently on top of ovary (e.g Umbelliferae)	P & A or K, C & A epigynous	Flower epigynous
fully inferior	(*h*)	K, C & A inserted on top of ovary, C & A fused (e.g. *Viburnum*)	K, C & A epigynous C & A borne on an epigynous zone	Flower epigynous
	(*i*)	K, C & A inserted on a ring or collar of tissue itself inserted on top of the ovary (e.g. *Fuchsia*)	K, C & A epigynous, C & A borne on an epigynous zone	Flower epigynous

The terms referring to ovary position are not especially ambiguous: a *superior* ovary is one borne on the receptacle (the apex of the pedicel) above the insertion of the other floral organs (regardless of whether these are free or variously united to each other), so that all these organs appear to arise from beneath the ovary. An *inferior* ovary is one borne below the point of insertion of the other floral organs (free or variously united) so that they appear to be borne on top of, or at least on the upper parts of the sides of, the ovary. A rather rare intermediate condition, in which the lower part of the ovary is inferior and the upper part superior (i.e. the other floral whorls appear to be borne about halfway up the ovary) is referred to as *half-inferior* or *semi-inferior*.

The position of the ovary is best determined by using a longitudinal section of the flower in the vertical plane (i.e. the plane in which the bract is borne). However, is it generally possible to tell whether the ovary is inferior by looking at the back of the flower, when it will be seen projecting below the base of the calyx.

The other terminology is more difficult to apply. One of the difficulties is that various authors have applied it to the whole of the flower (that is, the whole flower is described as hypogynous or perigynous or epigynous). This practice is misleading; the terms should be used to refer to the perianth and/or stamens only, following the original use of the terminology, as proposed by A. de Candolle in 1813. Table 1 and the accompanying illustrations (figs. 5, 6 and 7) should make clear the proper usage of these terms.

The terms may be difficult to use with unisexual flowers. If a *pistillode* (a sterile ovary) is present in a male flower it may be possible to decide whether the various whorls are hypogynous, perigynous or epigynous. If, however, no pistillode is present, much care must be taken and a female flower must be sought. If a purely female flower lacks a perianth (e.g. *Betula*) then it is not possible to use these terms at all, and such ovaries are described as *naked*.

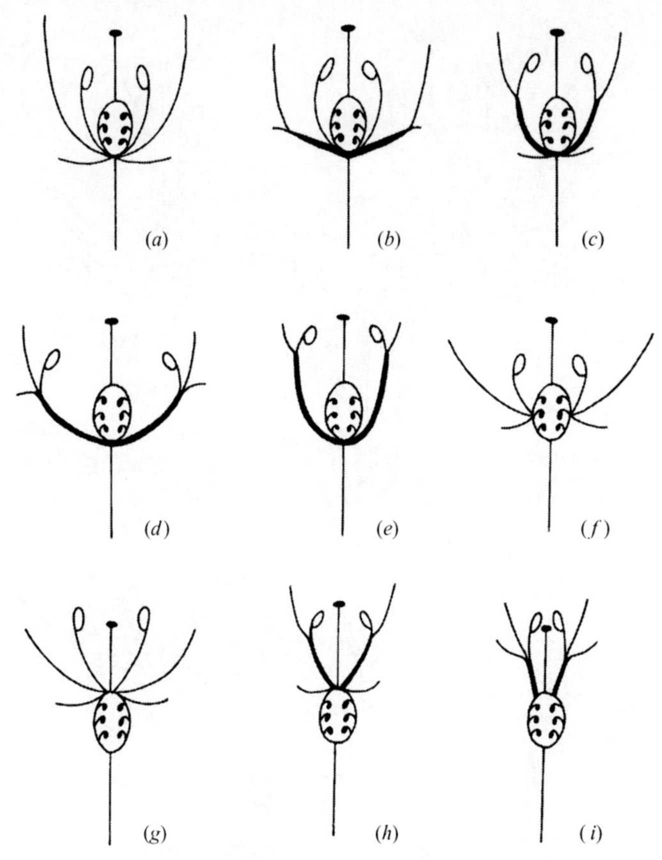

Figure 5. Diagrams illustrating the usage of the terms hypogyny, perigyny and epigyny. Perigynous and epigynous zones are indicated by the use of heavy lines. For further information see table 1.

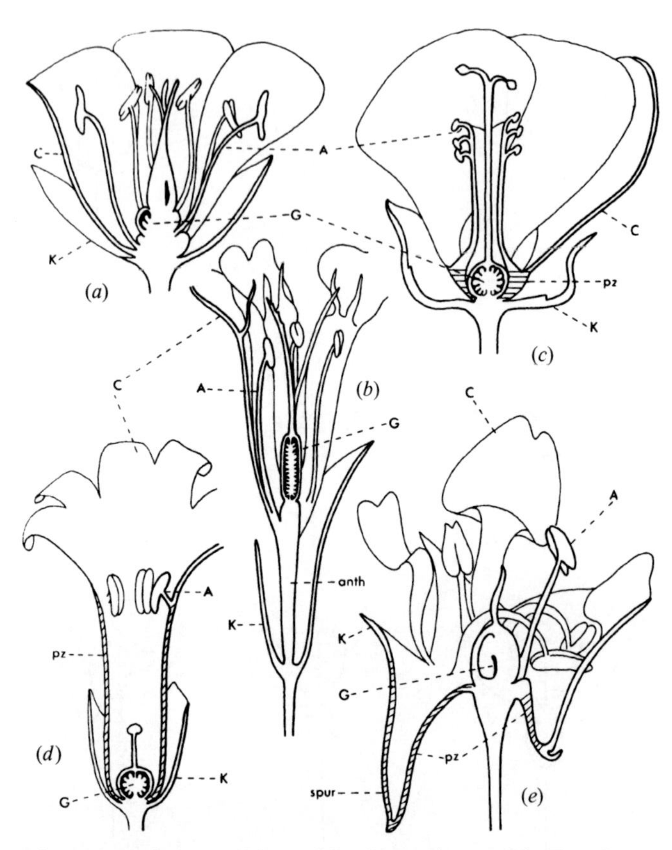

Figure 6. Relative positions of floral parts (see p. 37). Type I: (*a*) *Geranium*, (*b*) *Silene*. Type II: (*c*) *Abutilon*, (*d*) *Primula*. Type III: (*e*) *Tropaeolum*. A, androecium; anth, anthophore; C, corolla; G, gynoecium, K, calyx; pz, perigynous zone (hatched).

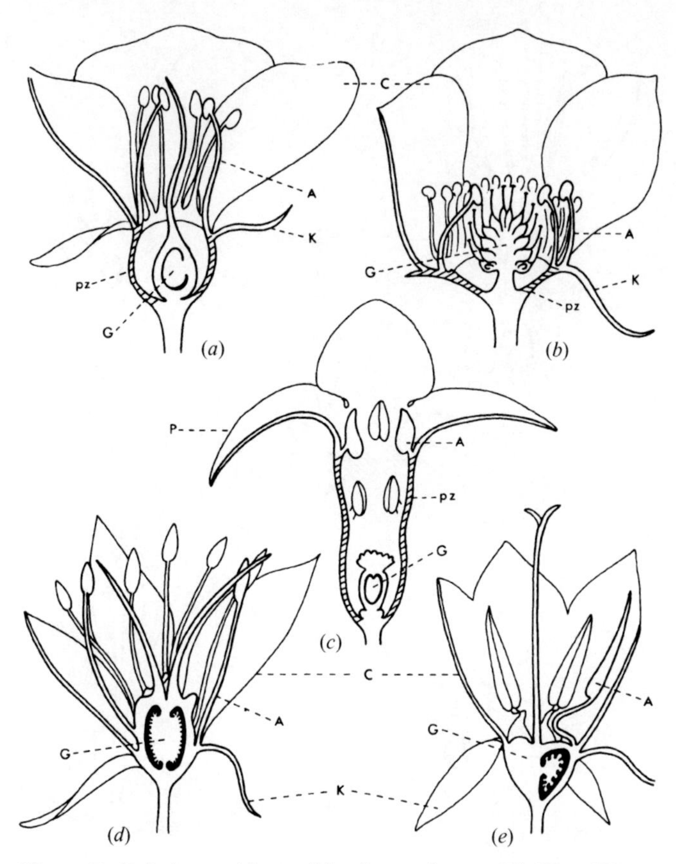

Figure 7. Relative positions of floral parts (see p. *37*). Type IV: (*a*) ***Prunus***, (*b*) ***Geum***, (*c*) ***Daphne***. Type V: (*d*) ***Saxifraga stolonifera***, (*e*) ***Campanula***. A, androecium; C, corolla; G, gynoecium; K, calyx; P, perianth (undifferentiated); pz, perigynous zone (hatched).

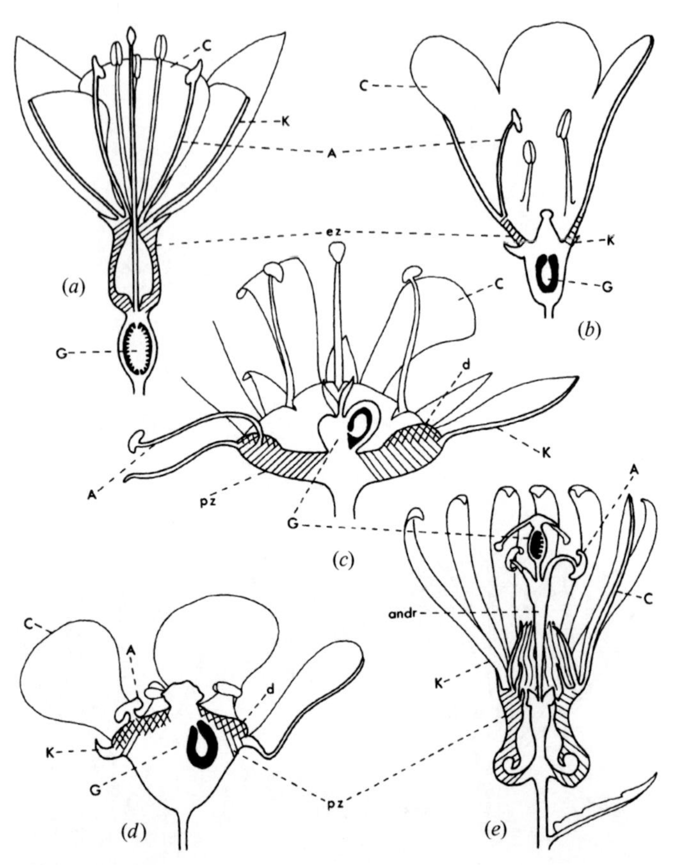

Figure 8. Relative positions of floral parts (see p. 37). Type VI: (*a*) *Fuchsia*, (*b*) *Viburnum*. More complicated types: (*c*) *Acer* (see p. XX), (*d*) *Euonymus* (see p. 45), (*e*) *Passiflora* (see p. 45). A, androecium; andr, androgynophore; C, corolla; d, disc (cross-hatched); ez, epigynous zone (hatched); G, gynoecium; K, calyx; pz, perigynous zone (hatched).

The following situations occur in the families covered by the key.

I Perianth and stamens hypogynous; ovary superior. See fig. 6*a,b*.
II Calyx hypogynous; corolla and stamens perigynous; ovary superior. See fig. 5*c,d*.
III Perianth (calyx and corolla) perigynous; stamens hypogynous; ovary superior. See fig. 7*e*.
IV Perianth (calyx and corolla) and stamens perigynous; ovary superior. See fig. 7*a–c*.
V Ovary partly or fully inferior; perianth (calyx and corolla) and stamens epigynous, without an epigynous zone. See fig. 7*d*.
VI Ovary partly or fully inferior; perianth (calyx and corolla) and stamens borne on an epigynous zone. See fig. 8*a,b*.

Types III and IV are often complicated by the presence of a *nectariferous disc* (a disc or ring of nectar-secreting tissue) surrounding the ovary, which may seem to be immersed in it. Usually, when such a disc is present, the perianth (as a clearly recognisable structure) is inserted on the edge of it. The stamens may also be borne on the edge of the disc, as in many species of maple (*Acer)* or on top of the disc, as in the spindle tree *(Euonymus)*. In the flowers of some other families the ovary can be borne on a long or short stalk (*gynophore*); this can be ignored when deciding whether the parts are hypogynous or perigynous (a gynophore is not possible when the parts are epigynous). A further complication arises in the passion flower (*Passiflora*) in which the ovary and stamens are borne on a common stalk (*androgynophore*) within the flowers; and in some members of the *Caryophyllaceae* (e.g. *Silene*) the corolla, stamens and ovary are borne on a short common stalk (known as an *anthophore*) within the flower.

The terminology as described above (hypogynous, perigynous, epigynous) is traditionally applied to the flowers of the dicotyledons only; the superior/inferior terminology is applied to both monocotyledons and dicotyledons (see p. 57 for the distinction between these large groups). In the descriptions of the families on pp. 125–259, the ovary position in the dicotyledons can be assumed from the description of the other parts; in the monocotyledons the ovary position is generally given for each family.

4 Fruits

In strict botanical terms, the fruit is the maturation-product of the ovary of a single flower. However, the term can be used more loosely to describe the structure that opens and/or releases the seed(s) for dispersal, or falls or is removed from the parent plant for the same purpose. This structure may simply be the fruit, as strictly defined above, or it may involve other parts of the former flower (sepals, styles, bracts, etc.) when it is known strictly as a *false fruit*, or the enclosed ovaries of several flowers, when it is known strictly as a *compound fruit*. The looser sense of the term will be used throughout this section.

In maturation into the fruit, the ovary wall becomes the fruit wall or *pericarp*. In many fruits the pericarp has a tripartite structure with a tough outer rind (the *exocarp*), a fleshy or fibrous central layer (the *mesocarp*) and a hard or stony inner layer (the *endocarp*), which surrounds the individual seed(s). The ovule(s) in the ovary mature into the individual seed(s), which are described in the next section.

There are many different kinds of fruit known in the flowering plants, and there is a very complex terminology used to describe them. Only those more frequently encountered will be included here. The criteria defining the different types of fruit include those mentioned above, and several others:

(a) whether the ovary from which the fruit was formed was apocarpous or syncarpous and superior or inferior;
(b) whether the fruit has a defined opening mechanism (a dehiscent fruit) or not (an *indehiscent* fruit);

(c) whether the fruit is fleshy or not, the fleshiness contributing to its dispersal by animals;
(d) how many seeds the fruit contains.

Follicle
A fruit derived from a single free carpel. Hence the actual fruit may be a single follicle (if the ovary from which it is derived was of a single carpel, e.g. *Consolida*) or a group of follicles (if the ovary was apocarpous and pluricarpellate, e.g. *Helleborus*). The follicle is dry, contains several seeds, and dehisces by opening out along its inner suture.

Legume
The legume or *pod* is similar to the follicle, but opens along both sutures. It is the characteristic fruit of the group of families formerly known as the Leguminosae (Fabaceae, Mimosaceae, Caesalpiniaceae).

Achene
A 1-seeded, dry, indehiscent fruit, which may be formed from an apocarpous or syncarpous, superior ovary (when formed from an apocarpous ovary the fruit is strictly a group of achenes, e.g. *Ranunculus*).

Cypsela
The equivalent of the achene when formed from an inferior ovary (e.g. all Compositae).

Caryopsis
Essentially an achene in which the pericarp and the seed coat (*testa*) have fused together, characteristic of the grass family (Gramineae).

Capsule
Formed from a pluricarpellate ovary, the capsule is dry, several-seeded, and dehisces by longitudinal (or rarely radial) splitting of

the pericarp, or by the formation of pores within the pericarp. It is the most common type of fruit.

Lomentum
The equivalent of a follicle, legume or capsule which breaks up into 1-seeded segments, which are themselves indehiscent. This type of fruit is found in the Fabaceae and Cruciferae and occasionally in other families.

Schizocarp
A dry fruit which splits into two or more generally 1-seeded, indehiscent parts (*mericarps*), which usually represent the carpels from which the ovary was formed (e.g. Umbelliferae).

Samara
An achene, cypsela or more rarely a mericarp that develops a conspicuous wing, which aids in its dispersal (e.g. *Acer*).

Nut
A large achene or mericarp with a very hard, woody pericarp. Smaller examples (e.g. *Polygonum*) are referred to as *nutlets*.

Berry
Essentially an indehiscent capsule in which the pericarp is fleshy and succulent.There may be one or more seeds (e.g. *Ribes*) or, if the endocarp becomes hard and bony around each seed, there may be one or several *stones* or *pyrenes*.

Drupe
An indehiscent fruit, usually containing a single seed, in which the pericarp is clearly 3-layered, with tough exocarp, fleshy mesocarp and hard endocarp. These are found especially in species of *Prunus* (cherry, plum, etc.). In *Rubus*, which has an apocarpous

ovary, the fruit consists of a group of partially united, small drupes, known as *drupelets* (raspberry, blackberry).

Pome

A fruit found only in those members of the Rosaceae that have inferior ovaries. Really a false fruit, involving the ovary and fleshy receptacular tissue which forms around it. The ovary inside the pome may be hard and stone-like (as in *Cotoneaster, Crataegus*, etc.) or parchment-like (e.g. *Malus, Pyrus*).

5 Seeds

The seeds are the maturation-products of the ovules and contain the potential for the next generation. Each contains an embryo (which may be rudimentary at dispersal, as is the case with orchids and many parasitic plants) and possible food-reserve material (*endosperm* or *perisperm*) wrapped inside the seed-coat (*testa*).

Seeds vary in size from minute and dust-like (as in many orchids) to large and solid. The testa may be winged or variously marked and coloured (it is shiny black in many families of monocotyledons) and the seed may be appendaged. Two kinds of appendage are important: the *aril*, which is an outgrowth from the *funicle* (the stalk of the ovule) and is often fleshy or coloured and may partially or wholly envelop the seed, and the *elaiosome* (also known as the *caruncle*), which is an oily body found at one end of the seed.

6 Miscellaneous features

Many plant parts have an *indumentum*, that is, they are covered with hairs of various forms. Hairs may be simple (unbranched), branched, *bifid* (with two arms, attached in the middle) or stellate, or modified into scale-like structures (*peltate scales*). Simple hairs may be unicellular or multicellular (often difficult to see without the aid of a microscope), and unbranched hairs may terminate in a glandular tip. There are many different terms used to describe hair covering, some of which (those used in the keys) are treated in the Glossary (pp. 269–283).

Most plants have clear sap, but in some families the sap is milky and/or coloured. This feature is most easily seen when the plant is growing rapidly and a leaf is broken from the stem. Care should be taken with the sap, which is sometimes irritant or poisonous.

Plants frequently bear spines on their stems or leaves; these are generally outgrowths of the stem and serve for protection from browsing animals, or as hooks for climbing or scrambling over other vegetation. Many climbers bear *tendrils*, fine, thread-like structures that coil around other vegetation or supports, enabling the plant to climb. Tendrils may be modified leaves, modified leaflets or modified inflorescences. In some cases the tendrils terminate in sticky pads, which enable the plant to grip the support more closely. In a few climbers (e.g. ivy, *Hedera*) specialised climbing roots grip the supports.

Very specialised leaves are found in insectivorous plants. These may be pitches or active traps. Botanical textbooks should be consulted for further details about these organs.

In some families the leaves, stems and/or petals contain glands or ducts containing aromatic oils. These can usually be seen as translucent or coloured dots or lines when the organ is held up to the light, and the oils can usually be smelt when the leaf itself is crushed (e.g. *Eucalyptus*).

Using the keys

The keys in this book are of the bracketed type and are dichotomous throughout, i.e. at every stage a choice must be made between two (and only two) contrasting alternatives (leads), which together make up a couplet. To facilitate reference to particular leads, each couplet is numbered and each lead is given a distinguishing letter (a or b). As the main key allows for the identification of over 320 families, it has been arranged into groups, with a key to the groups at the beginning.

To find the family to which a specimen belongs, one starts with the key to groups and compares the specimen with the two leads of the couplet numbered 1. If the specimen agrees with 1a, one proceeds to the lead with the number that is the same as that appearing at the right-hand end of lead 1a (in this case, 2); if, however, the plant agrees with lead 1b, then one proceeds to the couplet numbered 14. This process is repeated for subsequent couplets until, instead of a number at the right-hand end of a lead, a group is reached. Throughout this process, it is very important that the whole of each couplet be carefully read and understood before making a decision as to which lead to follow.

One proceeds in this way within the appropriate group key until the name of a family is reached. The families are numbered in the key; to provide a check on the identification obtained, the families are briefly described in numerical order on pp. 125–269. The specimen should be compared carefully with the description of the family: this should help to reveal errors made in observation or in the use of the key.

So that back-tracking should be easy, the number of the lead from which any particular couplet is derived is given in brackets after the couplet number of the 'a' lead. Thus '12a (3) . . .' means that one arrived at couplet 12 from one of the leads of couplet 3.

It will sometimes happen that the specimen will not agree with all the characters given in a particular lead. When this situation arises, one must decide which of the two leads of the relevant couplet the specimen agrees with more fully. In general, the most reliable diagnostics are put at the beginning of each lead, so these characters should be observed with particular care. The only exception to this occurs when the 'b' lead of a couplet reads: 'Combination of characters not as above'. In such cases the specimen must agree with *all* of the characters given in the 'a' lead; if it deviates in one or more characters it must be treated as falling into the 'combination of characters not as above' category (this strict interpretation may be tempered by the use of such words as 'sometimes' or 'usually' in the 'a' lead).

Many years' experience has shown that users make errors more often in the key to groups than anywhere else; of course, an error here means that a correct identification is virtually impossible. The following paragraphs form a commentary on the key to the groups, with precise indications of how it should be used. The principles adopted in this commentary, if not the details, are also relevant to the rest of the keys.

Couplet 1 discriminates between the dicotyledons and the monocotyledons, the two large groups into which the flowering plants (Angiospermae) naturally divide. There is no single character that completely and certainly distinguishes these two groups: instead, a combination of a rather large number of characters has to be used, and couplet 1 includes the most readily observed of these. 1a starts with the phrase 'Cotyledons usually 2, lateral', as opposed to 1b 'Cotyledon 1, terminal'. Specimens with two cotyledons (seedling leaves) clearly match with 1a, but those with only one cotyledon could fit either lead (the 'usually' in 1a indicates this), although it must be borne in mind that among the dicotyledons the occurrence of species with only a single cotyledon is extremely rare. Of course, the chances of being able to answer this question if the specimen is a mature plant are very small, but the character is important enough to be worth mentioning. The second phrases of the leads of the couplet are: 1a 'leaves usually net-veined, with or without stipules, alternate, opposite or whorled', as opposed to 1b 'leaves usually with parallel veins, sometimes these connected by cross-veinlets; leaves without stipules, opposite only in some aquatic plants'. These are admittedly imprecise alternatives, but they do help in distinguishing the two groups. Firstly, if the specimen has stipules (or the scars left by them after they have fallen), or if it is a terrestrial plant with opposite or whorled leaves, then clearly it matches with 1a. If the leaves are net-veined, then there is a high probability that it matches with 1a, just as if the leaves have parallel veins there is a high probability that it matches with 1b. The third characteristic used in this couplet reads: 1a 'flowers usually with parts in 2s, 4s or 5s or parts numerous' as opposed to 1b 'flowers usually with parts in 3s'. 'Parts' here essentially means sepals and petals or perianth and stamens. Again, this raises a matter of probabilities: if the flower has parts in 2s, 4s or 5s (or multiples of these), or the parts are numerous (more than 10), then there is a high probability that the plant belongs with 1a; of the parts are in 3s (or multiples), then there is a high probability that it belongs with 1b. Finally, the phrases 1a 'primary root system (taproot) usually persistent, branched', as opposed to 1b 'mature root system wholly adventitious' again pose a similar set of probabilities.

In making a decision about a particular plant, it is necessary to observe what information is available (both from the specimen, and from any other source) and then to make a judgement as to the balance of probabilities between 1a and 1b. Fortunately, with experience, the discrimination of these two large groups is actually easier than it might appear from the key, and little difficulty is generally experienced in deciding whether a particular plant belongs to one or the other.

Couplet 2 (arrived at if the plant matches 1a better than 1b) is much more straightforward. It asks whether the flower has distinct calyx and corolla, which are sharply distinguished from each other by position, size, texture or shape (matching 2a), or a perianth not distinguished into identifiable calyx and corolla (matching 2b).

Couplet 3, arrived at from 2a, is also very straightforward. If the ovary of the flower is partly or fully inferior, then 3a is appropriate; if the ovary is superior, then 3b should be taken. Couplet 4 (arrived at from 3a) is also simple, and leads to two distinct group keys. If the plant has flowers in which the petals are free from each other (or at least most of them are free from each other), then one proceeds to the Group I key; on the other hand, if all of the petals are united to each other below into a cup or a tube, then one proceeds to the Group II key.

If the plant matches 3b rather than 3a, then one proceeds to couplet 5. Here the distinction is similar to that found at couplet 4, but is worded somewhat differently to allow for the actual plants that fall under each lead. 5a includes plants with flowers in which most of the petals are free from each other, and fall individually, but it also includes some plants in which the petals are free from each other but are united at the base into a tube formed by the united filaments of the stamens, and therefore fall as a unit (corolla plus stamens), whereas 5b covers those plants in which the flower has a corolla made up of petals united at the base into a cup or tube.

Couplet 6, arrived at from 5a, separates those flowers with single or free carpels from those with united carpels. Note the wording about styles in each lead. 6a leads to Group III, whereas 6b continues to couplet 7.

Couplet 7 requires details of the number of stamens in relation to the number of petals. 7a covers those with stamens more than twice as many as petals (note that if the petals are 5, the stamens must be 11 or more to fit here) and leads to Group IV, whereas 7b includes those with stamens up to twice as many as petals (note that if the petals are 5, then the stamens in this group can be 10 or fewer).

Couplet 8 (arrived at from 7a) discriminates between different manners of placentation of the ovary; see pp. 25–31 for details.

Couplet 9 (arrived at from 8a) requires a decision on the arrangement of the leaves: alternate or reduced to alternate scales in 9a, opposite or whorled in 9b. Both of these lead to groups (Groups VI and VII).

Couplet 10 (arrived at from 5b) is concerned with the symmetry of the corolla. This can give rise to some problems; see the discussion of the matter on pp. 34–35.

Couplet 11, which is arrived at from 2b, discriminates those families in which the flowers are borne in catkins (Group X) from those in which the flowers are borne in any other manner.

Couplet 12 (arrived at from couplet 11b) is basically the same as couplet 6, except that it is here applied to plants whose flowers have a simple perianth.

Couplet 13, arrived at from couplet 12b, basically discriminates flowers with inferior ovaries (13a) and superior ovaries (13b). However, this is qualified by the phrases 'Stamens borne on the perianth' (13a) as opposed to 'Stamens free from the perianth' in an attempt to cover plants that have unisexual male flowers in which the ovary position cannot be determined (a special key that attempts to deal with such difficult male plants is provided on pp. 114–117).

Couplet 14, derived from 1b, deals with the monocotyledons, breaking them down into two large groups on the basis of whether the ovary is superior or inferior. The qualifying phrases in each couplet should be considered if the plant to be identified is aquatic.

Proceeding through the keys in this manner, making careful observations of the plant in the light of what is understood from a detailed reading of the various couplets, will lead to accurate identifications.

Keys

Key to Groups

See also the key to plant material that has only male flowers on p. 114.

1a Cotyledons usually 2, lateral; leaves usually net-veined, with or without stipules, alternate, opposite or whorled; flowers with parts in 2s, 4s or 5s, or parts numerous; primary root system (taproot) usually persistent, branched (Dicotyledons) **2**

b Cotyledon 1, terminal; leaves usually with parallel veins, sometimes these connected by cross-veinlets; leaves without stipules, opposite only in some aquatic plants; flowers usually with parts in 3s; mature root system wholly adventitious (Monocotyledons) **14**

2a (1) Perianth of 2 or rarely more whorls, distinguished usually into calyx and corolla, the outermost and inner whorls sharply distinguished by any or all of the following: position, colour, size, texture, shape **3**

b Perianth of a single whorl or rarely of 2 whorls that are not sharply distinguishable as above (there may be a relatively smooth transition from outer to inner), or completely absent **11**

3a (2) Ovary partly or fully inferior **4**

b Ovary totally superior **5**

4a (3) Most of the petals free from each other at the base **Group I (p. 59)**

b All petals united into a tube or cup at the base **Group II (p. 62)**

5a (3) Corolla made up of petals at least some of which are free from each other at their bases, falling individually except rarely when either attached individually to a ring formed by the united bases of the filaments or joined loosely at the apex **6**
b All petals united into a tube at the base **10**
6a (5) Ovary of a single carpel with a single style and/or stigma, or made up of several carpels that are entirely free from each other (including their styles) **Group III (p. 64)**
b Ovary of 2 or more carpels that are united to each other at least by their styles, more usually the bodies of the carpels united **7**
7a (6) Stamens more than twice as many as petals **Group IV (p. 67)**
b Stamens up to twice as many as petals **8**
8a (7) Placentation parietal **Group V (p. 71)**
b Placentation axile, apical, basal or free-central **9**
9a (8) Leaves alternate, or reduced to alternate scales **Group VI (p. 73)**
b Leaves opposite or whorled **Group VII (p. 79)**
10a (5) Corolla actinomorphic **Group VIII (p. 81)**
b Corolla zygomorphic **Group IX (p. 87)**
11a (2) At least the male flowers borne in catkins, which are usually deciduous as a whole **Group X (p. 90)**
b Flowers not borne in catkins as above **12**
12a (11) Ovary of a single carpel with a single style and/or stigma, or made up of several carpels that are entirely free from each other (including their styles) **Group XI (p. 92)**
b Ovary of 2 or more carpels that are united to each other at least by their styles, more usually the bodies of the carpels united **13**
13a (12) Stamens borne on the perianth, or ovary inferior **Group XII (p. 95)**
b Stamens free from the perianth, ovary superior **Group XIII (p. 98)**

14a (1) Ovary superior or flowers completely without perianth (including all aquatics with totally submerged flowers) **Group XIV (p. 102)**
b Ovary partly or fully inferior (if plants aquatic then flowers borne above water level) **Group XV (p. 110)**

Group I

Dicotyledons with perianth of 2 distinct whorls (calyx and corolla); ovary partly or fully (inferior; petals) free from each other.

1a Petals and stamens numerous; plants succulent 2
b Petals 10 or fewer, stamens usually fewer than 10; plants usually not succulent 3
2a (1) Stems succulent, usually with spines; leaves usually absent **26 Cactaceae**
b Leaves succulent; spines usually absent **19 Aizoaceae**
3a (1) Anthers opening by terminal pores 4
b Anthers opening by longitudinal slits or by valves 5
4a (3) Filaments each with a knee-like joint below the anther; leaves usually with 3 conspicuous main veins from the base **150 Melastomataceae**
b Filaments straight; leaves each with a single main vein **151 Rhizophoraceae**
5a (3) Placentation parietal, placentas sometimes intrusive 6
b Placentation axile, apical, basal or free-central 10
6a (5) Leaves with translucent, aromatic glands **147 Myrtaceae**
b Leaves without translucent, aromatic glands 7
7a (5) Aquatic plants with large, floating, peltate leaves **45 Nymphaeaceae**
b Combination of characters not as above 8
8a (7) Stamens 8 or more; leaves usually opposite **76c Hydrangeaceae**

b Stamens 4–6; leaves alternate 9

9a (8) Disc present; leaves usually with gland-tipped teeth **76d Escalloniaceae**

b Disc absent; leaves without gland-tipped teeth **76e Grossulariaceae**

10a (5) Placentation free-central; sepals 2 **20 Portulacaceae**

b Placentation axile, apical or basal; sepals usually more than 2 11

11a (10) Stamens as many as and on the same radii as petals; trees or shrubs with simple leaves **120 Rhamnaceae**

b Stamens more numerous than petals or if as many, then not on the same radii as them; plants herbaceous or woody, leaves simple or compound 12

12a (11) Leaves with translucent, aromatic glands **147 Myrtaceae**

b Leaves without translucent, aromatic glands 13

13a (12) Style 1 14

b Styles 2 to numerous 27

14a (13) Floating aquatic herbs with inflated leaf-stalks **146 Trapaceae**

b Terrestrial herbs, trees or shrubs; leaf-stalks not inflated 15

15a (14) Small, low shrubs with scale-like, overlapping leaves; flowers in axillary heads **81 Bruniaceae**

b Combination of characters not as above 16

16a (15) Inflorescences borne on the surfaces of the leaves (by adnation of the peduncle to the leaf main vein) **161d Helwingiaceae**

b Inflorescences not borne on the leaf surfaces 17

17a (16) Ovule 1, apical in each cell of the ovary (the ovary may be 1-celled) 18

b Ovules 2 to many in each cell of the ovary (the ovary may be 1-celled) 24

18a (17) Stamens with swollen, hairy filaments; petals rolled and recurved downwards **158 Alangiaceae**

b Stamens without swollen, hairy filaments; petals often borne horizontally, but not as above **19**
19a (18) Ovary with 2 or more cells **20**
b Ovary single-celled **21**
20a (19) Sepals 5, petals 5 **161a Torricelliaceae**
b Sepals 4, petals 4 **161 Cornaceae**
21a (19) Petals 5 (or rarely more), imbricate **22**
b Petals 4, valvate **23**
22a (21) Stigmas 3; leaves evergreen **161c Griseliniaceae**
b Stigmas 2; leaves deciduous **159 Nyssaceae**
23a (21) Flowers unisexual; petals brownish; leaves evergreen **161b Aucubaceae**
b Flowers bisexual; petals various, not brownish; leaves usually deciduous **161 Cornaceae**
24a (17) Stamens more than 10; ovary with 8–12 superposed cells; plant a spiny shrub **148 Punicaceae**
b Combination of characters not as above **25**
25a (24) Stamens 8–10; plants woody **152 Combretaceae**
b Stamens 4–8; plants herbaceous **26**
26a (25) Sap watery; petals 2 or 4; ovary usually 4-celled **153 Onagraceae**
b Sap milky; petals 5; ovary 3-celled **215 Campanulaceae**
27a (13) Flowers borne in umbels, these sometimes modified, or in superposed whorls; leaves usually compound or much divided **28**
b Flowers not borne in umbels; leaves usually simple, little divided **29**
28a (27) Fruit a schizocarp splitting into 2 mericarps; flowers usually bisexual; petals imbricate in bud and inflexed; usually aromatic herbs without stellate hairs **164 Umbelliferae**
b Fruit a berry; flowers often unisexual; petals valvate in bud, not inflexed; plants mostly woody, often with stellate hairs **163 Araliaceae**

29a (27) Plants herbaceous **30**
b Plants woody **31**
30a (29) Leaves deeply dissected; stamens usually 8; ovules 1–4, apical **154 Haloragaceae**
b Leaves not as above; stamens usually 10; ovules numerous, axile **76 Saxifragaceae**
31a (29) Anthers opening by valves; stellate hairs often present **73 Hamamelidaceae**
b Anthers opening by slits; stellate hairs absent **32**
32a (31) Leaves opposite, evergreen **77 Cunoniaceae**
b Leaves mainly alternate and deciduous, never both evergreen and opposite **82 Rosaceae**

Group II

Dicotyledons with perianth of 2 distinct whorls (calyx and corolla); ovary partly or fully inferior; petals united to each other at the base.

1a Leaves needle-like or scale-like; small, heather-like shrublets **81 Bruniaceae**
b Combination of characters not as above **2**
2a (1) Leaves whorled, mostly basal, leathery, spiny; inflorescence a spike of many-flowered whorls; calyx 2-lobed **214a Morinaceae**
b Combination of characters not as above **3**
3a (2) Inflorescence a head surrounded by an involucre of bracts; ovule always solitary **4**
b Inflorescence and ovules not as above **6**
4a (3) Each flower with a cup-like involucel; anthers not united into a tube around the style **214 Dipsacaceae**
b Involucel absent; anthers united into a tube around the style **5**
5a (4) Ovule basal; calyx represented by a variable number of scales or hairs (pappus) or effectively absent **219 Compositae**

b Ovule apical; calyx of 4–6 teeth **218a Calyceraceae**

6a (3) Stamens 2, united to the style to form a touch-sensitive column; leaves linear **218 Stylidiaceae**

b Combination of characters not as above **7**

7a (6) Leaves alternate or all basal **8**

b Leaves opposite or whorled **17**

8a (7) Anthers opening by pores; fruit a berry or drupe **168 Ericaceae**

b Anthers opening by longitudinal slits; fruit various **9**

9a (8) Evergreen trees or shrubs; corolla white, campanulate; ovary half-inferior; placentation free-central, ovules few **172 Myrsinaceae**

b Combination of characters not as above **10**

10a (9) Climbers with tendrils and unisexual flowers; stamens 1–5; placentation parietal; fruit berry-like **144 Cucurbitaceae**

b Combination of characters not as above **11**

11a (10) Stamens 10 to many; plants woody **12**

b Stamens fewer than 6; plants woody or herbaceous **14**

12a (11) Leaves with translucent glands smelling of eucalyptus; corolla completely united, unlobed, falling as a whole **147 Myrtaceae**

b Combination of characters not as above **13**

13a (12) Hairs stellate or scale-like; stamens in 1 series, anthers linear **177 Styracaceae**

b Hairs absent or not as above; stamens in several series; anthers broad **178 Symplocaceae**

14a (11) Stigmas surrounded by a sheath formed from the top of the style **216 Goodeniaceae**

b Stigmas not surrounded by a sheath **15**

15a (14) Stamens as many as, and on the same radii as, the petals **173 Primulaceae**

b Stamens not as above **16**

16a (15) Stamens 2 or 4, borne on the corolla; sap not milky **205 Gesneriaceae**

b Stamens 5, free from the corolla; sap usually milky
215 Campanulaceae

17a (7) Placentation parietal; stamens 2, or 4 and paired
205 Gesneriaceae

b Placentation axile or apical; stamens 1 or more, if 4 then not paired **18**

18a (17) Stamens 1–3; ovary with a single ovule
213 Valerianaceae

b Stamens 4 or 5; ovary with usually 2 or more ovules **19**

19a (18) Leaves divided into 3 leaflets; flowers few, in a head; herbaceous **212 Adoxaceae**

b Leaves simple or rarely pinnate; inflorescence various, usually not as above; usually woody **20**

20a (19) Stipules usually borne between the bases of the leaf-stalks and sometimes looking like leaves; ovary usually 2-celled, more rarely 5-celled; corolla usually actinomorphic; fruit capsular, fleshy or schizocarpic **186 Rubiaceae**

b Stipules usually absent, when present not as above; ovary usually 3-celled (occasionally 2–5-celled), sometimes only 1 cell fertile; corollas often zygomorphic; fruit a berry or drupe
211 Caprifoliaceae

Group III

Dicotyledons with perianth of 2 distinct whorls (calyx and corolla); ovary superior; petals free from each other; ovary of a single carpel or several free carpels.

1a Ovary apparently consisting of a single carpel, with a single style and/or stigma and a singe cell within, with 1 to many ovules **2**

b Ovary consisting of 2 or more carpels that are entirely free from each other, each with its own separate style and stigma **8**

2a (1) Leaves often pinnate, bipinnate, trifoliolate or palmate, rarely simple or reduced to phyllodes, with stipules **3** b Leaves often simple, without stipules 5

3a (2) Corolla actinomorphic, petals free or somewhat united at the base, they or the lobes valvate in bud; stamens 4 to many; leaves bipinnate, rarely reduced to phyllodes; seeds each with a U-shaped lateral line **84b Mimosaceae**

b Corolla zygomorphic; petals mostly free (sometimes 2 of them united), imbricate in bud; stamens 10 or fewer; seeds usually without a lateral line, rarely with a closed lateral line 4

4a (3) Upper petal interior (rarely petal 1 or petals absent); seed usually with a straight radicle **84a Caesalpiniaceae**

b Upper petal exterior; seed usually with an incurved radicle **84 Fabaceae**

5a (2) Corolla actinomorphic 6

b Corolla zygomorphic 7

6a (5) Resinous tree or shrub; style set obliquely on the ovary **105 Anacardiaceae**

b Non-resinous shrubs or herbs; style not set obliquely on the ovary **42 Berberidaceae**

7a (5) Stamens 3 or 4, anthers opening by pores **85 Krameriaceae**

b Stamens more numerous, anthers opening by slits **41 Ranunculaceae**

8a (1) Calyx, corolla and stamens perigynous 9

b Calyx, corolla and stamens hypogynous 10

9a (8) Stipules absent; leaves entire; flowers solitary and terminal; seeds each with a divided aril **54 Crossosomataceae**

b Stipules present; leaves usually toothed, lobed or divided (if entire then flowers clustered); seeds without arils **82 Rosaceae**

10a (8) Aquatic plants with floating or emergent peltate leaves (submerged leaves may be of different shape) 11

b Terrestrial plants, no leaves peltate 12

11a (10) Carpels sunk individually in a top-shaped receptacle; sepals 4–5, petals 10–25 **45b Nelumbonaceae**

b Carpels not sunk in a receptacle; sepals 3, petals 3 **45a Cabombaceae**
12a (10) Leaves conspicuously succulent **74 Crassulaceae**
b Leaves not succulent 13
13a (12) Plants completely herbaceous 14
b Plants woody 17
14a (13) Petals fringed; fruits borne on a common gynophore **69 Resedaceae**
b Petals not fringed; gynophore absent 15
15a (14) Sap milky **66 Papaveraceae**
b Sap clear, watery 16
16a (15) Sepals not all the same size and shape; stamens borne on a nectar-secreting disc **53 Paeoniaceae**
b Sepals all similar in shape and size; stamens not borne on a disc, nectar secreted on the petals **41 Ranunculaceae**
17a (13) Leaves opposite; each petal keeled inside **104 Coriariaceae**
b Leaves alternate; petals not keeled inside 18
18a (17) Leaves simple, entire or toothed 19
b Leaves compound or deeply lobed or divided 22
19a (18) Woody climbers with unisexual flowers; petals 3, 6 or 9 20
b Shrubs; flowers not as above 21
20a Stamens united into a fleshy mass; ovules 2–3 per carpel **32 Schisandraceae**
b Stamens free; ovules 1 per carpel **44 Menispermaceae**
21a (18) Leaves dotted with translucent glands; petals in 2 or more series **28 Winteraceae**
b Leaves without translucent glands; petals in a single whorl **52 Dilleniaceae**
22a (18) Flowers unisexual; mostly woody climbers; if shrubs, then with blue fruits **43 Lardizabalaceae**
b Flowers bisexual; shrubs, fruits never blue 23

23a (22) Sepals not all the same size and shape; stamens borne on a nectar-secreting disc **53 Paeoniaceae**
b Sepals all similar in size and shape; stamens not borne on a disc, nectar secreted on the petals **41 Ranunculaceae**

Group IV

Dicotyledons with distinct calyx and corolla; petals free from each other at their bases; ovary of 2 or more united carpels; stamens more than twice as many as petals.

1a Herbaceous climber; leaves palmately divided into stalked leaflets; petals 2, stamens 8 **90 Tropaeolaceae**
b Combination of characters not as above 2
2a (1) Perianth and stamens hypogynous, borne independently below the superior ovary 3
b Perianth and stamens perigynous, borne on the edge of a rim or cup, which itself is borne below the superior ovary 34
3a (2) Placentation axile or free-central 4
b Placentation parietal 23
4a (3) Placentation free-central; sepals 2 **20 Portulacaceae**
b Placentation axile; sepals usually more than 2 5
5a (4) Leaves all basal, tubular, forming insect-trapping pitchers; style peltately dilated **63 Sarraceniaceae**
b Leaves not as above; style not peltately dilated 6
6a (5) Leaves alternate 7
b Leaves opposite or rarely whorled 20
7a (6) Anthers opening by terminal pores 8
b Anthers opening by longitudinal slits 10
8a (7) Shrubs with simple leaves without stipules, often covered with stellate hairs; stamens inflexed in bud; fruit a berry **56 Actinidiaceae**
b Combination of characters not as above 9

9a (8) Ovary deeply lobed, borne on an enlarged receptacle or gynophore; petals not fringed **57 Ochnaceae**
b Ovary not lobed, not borne as above; petals often fringed **123 Elaeocarpaceae**
10a (7) Perianth segments of inner whorl (petals) tubular or bifid, nectar-secreting; fruit a group of partly to fully coalescent follicles **41 Ranunculaceae**
b Combination of characters not as above 11
11a (10) Leaves with translucent, aromatic glands **96 Rutaceae**
b Leaves without such glands 12
12a (11) Sap milky; flowers unisexual **94 Euphorbiaceae**
b Sap watery; flowers bisexual 13
13a (12) Succulent herb with spines; bark hard and resinous; stamens 15 in groups of 3 in each of which the central is the largest **89 Geraniaceae**
b Combination of characters not as above 14
14a (13) Large tropical trees; sepals 5, all or 2 or 3 of them enlarged and wing-like in fruit **58 Dipterocarpaceae**
b Combination of characters not as above 15
15a (14) Stipules absent; leaves evergreen **59 Theaceae**
b Stipules present; leaves usually deciduous 16
16a (15) Filaments free; anthers 2-celled 17
b Filaments united into a tube at least around the ovary, often also around the style; anthers often 1-celled 18
17a (16) Nectar-secreting disc absent; stamens more than 15; leaves simple **124 Tiliaceae**
b Nectar-secreting disc present, conspicuous; stamens 15; leaves dissected **91 Zygophyllaceae**
18a (16) Styles divided above, several; stipules often persistent; carpels 5 or more **125 Malvaceae**
b Style 1, stigma capitate or several; stipules usually deciduous; carpels 2–5 19
19a (18) Stamens in 2 whorls, those of the outer whorl usually sterile **127 Sterculiaceae**

b Stamens in several whorls, all fertile **126 Bombacaceae**

20a (6) Sepals united, falling as a unit; fruit separating into two boat-shaped units **55 Eucryphiaceae**

b Sepals and fruit not as above 21

21a (20) Small trees; stamens with brightly coloured filaments which are at least twice as long as the petals, the anthers forming a circle **60 Caryocaraceae**

b Combination of characters not as above 22

22a (21) Leaves simple, without stipules, often with translucent glands; stamens often united in bundles **62 Guttiferae**

b Leaves pinnate, without translucent glands; stamens not united in bundles **91 Zygophyllaceae**

23a (3) Aquatic plants with cordate leaves; style and stigmas forming a disc on top of the ovary **45 Nymphaeaceae**

b Combination of characters not as above 24

24a (23) Leaves modified into active insect-traps, the 2 halves of the blade fringed and closing rapidly when stimulated **65 Droseraceae**

b Leaves not as above 25

25a (24) Leaves opposite 26

b Leaves alternate 28

26a (25) Styles numerous; floral parts in 3s **66 Papaveraceae**

b Styles 1–5; floral parts in 4s or 5s 27

27a (26) Style 1; stamens not united in bundles; leaves without translucent glands **135 Cistaceae**

b Styles 3–5, free or variously united below; stamens united in bundles (rarely apparently all free); leaves with translucent or blackish glands **62 Guttiferae**

28a (25) Small trees with aromatic bark; filaments of the stamens all united **31 Canellaceae**

b Herbs shrubs or trees, bark not aromatic; filaments free 29

29a (28) Trees; leaves with stipules; anthers opening by short, pore-like slits 30 b Herbs or shrubs; leaves usually without stipules; anthers opening by longitudinal slits 31

30a (29) Anthers horseshoe-shaped; leaves simple, entire **136 Bixaceae**

b Anthers straight; leaves palmately lobed **136a Cochlospermaceae**

31a (29) Sepals 2 or rarely 3, quickly deciduous **66 Papaveraceae**

b Sepals 4–8, persistent in flower 32

32a (31) Leaves scale-like; styles 5, stigmas 5 **137 Tamaricaceae**

b Leaves not as above; styles 1, 2, 3 or absent, stigmas 1, 2 or 3 33

33a (32) Ovary closed at the apex, borne on a stalk (gynophore); none of the petals fringed **67 Capparaceae**

b Ovary open at the apex, not borne on a stalk; at least some of the petals fringed **69 Resedaceae**

34a (2) Flowers unisexual; leaf-bases oblique **143 Begoniaceae**

b Flowers bisexual; leaf-bases not oblique 35

35a (34) Filaments each with a knee-like joint below the anther; leaves usually with 3 conspicuous main veins from the base **150 Melastomataceae**

b Filaments straight; leaves each with a single main vein 36

36a (35) Placentation free-central; ovary partly inferior **20 Portulacaceae**

b Placentation not free-central; ovary either completely superior or completely inferior 37

37a (36) Aquatic plants with cordate leaves **45 Nymphaeaceae**

b Terrestrial plants; leaves various 38

38a (37) Carpels 1 or 3, eccentrically placed at the top of, the bottom of, or within the tubular perigynous zone **83 Chrysobalanaceae**

b Carpels and perigynous zone not as above 39

39a (38) Stamens united into bundles on the same radii as the petals; staminodes often present; plants usually rough with stinging hairs **141 Loasaceae**

b Combination of characters not as above 40

40a (39) Sepals 2, united, falling as a unit as the flower opens; plants herbaceous **66 Papaveraceae**
b Sepals 4 or 5, usually free, not falling as a unit; mostly trees or shrubs 41
41a (40) Stamens united into several rings or sheets **149 Lecythidaceae**
b Stamens not as above 42
42a (41) Carpels 8–12, superposed **145 Lythraceae**
b Carpels fewer, side-by-side 43
43a (42) Leaves with stipules 44
b Leaves without stipules 46
44a (43) Leaves alternate; plants woody or herbaceous **82 Rosaceae**
b Leaves opposite; plants woody 45
45a (44) Leaves simple; anthers opening by short, pore-like slits **123 Elaeocarpaceae**
b Leaves compound; anthers opening by full-length slits **77 Cunoniaceae**
45a (44) Leaves with translucent, aromatic glands; style 1 **147 Myrtaceae**
b Leaves without such glands; styles more than 1 **76c Hydrangeaceae**

Group V

Dicotyledons with calyx and corolla; petals free from each other at the base; stamens up to twice as many as petals; ovary of 2 or more united carpels; placentation parietal.

1a Sepals, petals and stamens perigynous, borne on a rim or cup, which itself is inserted below the ovary 2
b Sepals, petals and stamens hypogynous, inserted individually below the ovary 7
2a (1) Trees; leaves bi- or tripinnate; flowers bilaterally symmetric; stamens 5, of different lengths **70 Moringaceae**

b Combination of characters not as above 3
3a (2) Annual aquatic herb; stamens 6 **68 Cruciferae**
b Combination of characters not as above 4
4a (3) Flower-stalks slightly united to the leaf-stalks so that the flowers appear to be borne on the latter; petals contorted in bud; carpels 3 **133 Turneraceae**
b Flower-stalks not united to the leaf-stalks; petals not contorted in bud; carpels usually 2 or 4 5
5a (4) Stamens 4–6 **76d Escalloniaceae**
b Stamens 8 or more 6
6a (5) Ovary surrounded by a disc bearing 10 small staminode-like structures; placentas 5, very intrusive **110a Greyiaceae**
b Disc absent, without staminodes; placentas 2–4, not intrusive **76c Hydrangeaceae**
7a (1) Corolla zygomorphic 8
b Corolla actinomorphic 11
8a (7) Ovary open at apex; some or all petals fringed **69 Resedaceae**
b Ovary closed at the apex; no petals fringed 9
9a (8) Petals and stamens 5; carpels 2 or 3 **131 Violaceae**
b Petals and stamens 4 or 6; carpels 2 10
10a (9) Ovary borne on a stalk (gynophore); stamens projecting well beyond the petals **67 Capparaceae**
b Ovary not borne on a stalk; stamens not projecting beyond petals **66a Fumariaceae**
11a (7) Plants parasitic, without chlorophyll; anthers each opening by a single horseshoe-shaped slit **168a Monotropaceae**
b Plants free-living, with chlorophyll; anthers usually opening by longitudinal slits, never as above 12
12a (11) Petals and stamens numerous **19 Aizoaceae**
b Petals and fertile stamens each fewer than 10 13
13a (12) Stamens alternating with much-divided staminodes **76b Parnassiaceae**
b Stamens not alternating with much-divided staminodes 14

14a (13) Leaves insect-trapping and -digesting by means of stalked, glandular hairs **65 Droseraceae**
b Leaves not as above 15
15a (14) Climbers 16
b Shrubs or herbaceous plants 17
16a (15) Plants with tendrils; ovary and stamens borne on a common stalk (androgynophore); corona present **134 Passifloraceae**
b Plant without tendrils; ovary and stamens not borne on a common stalk; corona absent **130 Flacourtiaceae**
17a (15) Petals 4, the outer pair trifid; sepals 2 **66a Fumariaceae**
b Petals not as above; sepals 4 or 5 18
18a (17) Stamens usually 6, 4 longer and 2 shorter, rarely reduced to 2; carpels 2; fruit usually with a secondary septum **68 Cruciferae**
b Stamens 4–10, all more or less equal; carpels 2–5; fruit without a secondary septum 19
19a (18) Petals each with a scale-like appendage at the base of the blade; leaves opposite **138 Frankeniaceae**
b Petals without appendages; leaves alternate or all basal 20
20a (19) Stipules present **131 Violaceae**
b Stipules absent 21
21a (20) Leaves alternate, scale-like **137 Tamaricaceae**
b Leaves usually all basal, not scale-like **167 Pyrolaceae**

Group VI

Dicotyledons with calyx and corolla present; petals free from each other; ovary of united carpels; stamens up to twice as many as petals; placentation axile, basal, apical or free-central; leaves alternate.

1a Placentation free-central; ovary of a single cell, at least above 2
b Placentation axile, basal or apical; ovary of a single cell or of 2 or more cells 4

2a (1) Shrubs; leaves mostly evergreen with translucent dots or stripes; style 1 **172 Myrsinaceae**
b Combination of characters not as above **3**
3a (2) Sepals usually 2; if more, then petals numerous **20 Portulacaceae**
b Sepals 4 or 5; petals 4 or 5 **22 Caryophyllaceae**
4a (1) Stamens (including staminodes) and petals usually of the same number and on the same radii (stamens antepetalous), rarely stamens fewer than petals **5**
b Stamens not on the same radii as the petals **11**
5a (4) Styles 5, free or shortly joined towards the base; ovule 1, basal, borne on a long, curved funicle **174 Plumbaginaceae**
b Combination of characters not as above **6**
6a (5) Fertile stamens 2, staminodes 3; corolla zygomorphic **109a Meliosmaceae**
b All stamens (4 or 5) fertile; corolla actinomorphic **7**
7a (6) Sepals, petals and stamens perigynous **120 Rhamnaceae**
b Sepals, petals and stamens hypogynous **8**
8a (7) Inflorescences leaf-opposed; climbers with tendrils, or rarely shrubs **9**
b Inflorescences not leaf-opposed; usually trees **10**
9a (8) Filaments of stamens free from each other at the base **121 Vitaceae**
b Filaments of stamens united to each other at the base **122 Leeaceae**
10a (8) Leaves with stipules, evergreen **114 Corynocarpaceae**
b Leaves without stipules, usually deciduous **109 Sabiaceae**
11a (4) Anthers opening by clearly defined pores at the apex **12**
b Anthers opening by longitudinal or horseshoe-shaped slits or by valves **19**
12a (11) Leaves and stems covered in conspicuous glandular hairs on which insects are often trapped **13**
b Leaves and stems without such hairs **14**

13a (12) Carpels 2; herbs **79 Byblidaceae**
b Carpels 3; low shrubs **80 Roridulaceae**
14a (12) Low shrubs with unisexual flowers; stamens 4, petals 4, some of them 2–3-lobed **123 Elaeocarpaceae**
b Combination of characters not as above **15**
15a (14) Corolla zygomorphic; stamens 8 **103 Polygalaceae**
b Corolla actinomorphic; stamens some other number **16**
16a (15) Carpels 2; leaves opposite **102 Tremandraceae**
b Carpels 3 or more; leaves alternate **17**
17a (16) Carpels 3; style divided above into 3 stigmas **166 Clethraceae**
b Carpels 4 or more; style undivided or with 4 or more branches **18**
18a (17) Petals about as broad as long, clawed; evergreen herbs or low shrubs; style divided above into 4 or 5 stigmas, rarely unlobed **167 Pyrolaceae**
b Petals longer than broad; styles undivided, stigmas 4 or 5, borne in a cup-like sheath **168 Ericaceae**
19a (11) Corolla zygomorphic **20**
b Corolla actinomorphic **25**
20a (19) Anthers cohering above the ovary like a cap **111 Balsaminaceae**
b Anthers not cohering as above **21**
21a (20) Leaves with stipules **22**
b Leaves without stipules **23**
22a (21) Stamens 4, free; stipules borne between the petioles and the stems **110 Melianthaceae**
b Stamens 10 or more, filaments united into a tube around the styles; stipules borne laterally to the petioles **89 Geraniaceae**
23a (21) Stamens 8; sprawling or climbing herbs; leaves peltate or variously divided **90 Tropaeolaceae**
b Combination of characters not as above **24**
24a (23) Plants herbaceous **76 Saxifragaceae**

b Plants woody **107 Sapindaceae**
25a (19) Sepals, petals and stamens perigynous **26**
b Sepals, petals and stamens hypogynous **30**
26a (25) Style 1, often divided above **27**
b Styles more than 1, often 2 and divergent **28**
27a (26) Herbs; stipules present; fruit a nut distributed within the persistent calyx **22a Illecebraceae**
b Shrubs or trees; stipules usually absent; fruit generally a several-seeded capsule **115 Celastraceae**
28a (26) Fruit an inflated, membranous capsule; leaves compound **116 Staphyleaceae**
b Fruit not as above; leaves simple **29**
29a (28) Trees or shrubs; hairs often stellate; anthers usually opening by valves; fruit a few-seeded, woody capsule **73 Hamamelidaceae**
b Herbs; hairs simple or absent; fruit a capsule or almost a pair of separate follicles **76 Saxifragaceae**
30a (25) Petals and stamens both 8 or more; stamens numerous **31**
b Petals and stamens fewer than 8; stamens usually definite in number **33**
31a (30) Aquatic herbs with floating leaves **45 Nymphaeaceae**
b Land plants often from desert areas **32**
32a (31) Annual herb covered with stellate hairs **17c Molluginaceae**
b Usually perennial herbs, hairless or with simple hairs **19 Aizoaceae**
33a (30) Leaves with translucent, aromatic glands **96 Rutaceae**
b Leaves without such glands **34**
34a (33) Sap usually milky; flowers unisexual; styles 3, often further divided **94 Euphorbiaceae**
b Combination of characters not as above **35**
35a (34) Plant parasitic, without chlorophyll; anthers each opening by a single horseshoe-shaped slit **168a Monotropaceae**

b Plant free-living, with chlorophyll; anthers opening by usually 2 longitudinal slits **36**

36a (35) Flower with a well-developed nectar-secreting disc below and around the ovary **37**

b Disc absent, nectar secreted in other ways **44**

37a (36) Resinous trees or shrubs **38**

b Herbs, shrubs or trees, not resinous, occasionally aromatic **39**

38a (37) Ovules 2 in each cell of the ovary **99 Burseraceae**

b Ovule 1 in each cell of the ovary **105 Anacardiaceae**

39a (37) Leaves reduced to overlapping scales **137 Tamaricaceae**

b Leaves well developed **40**

40a (39) Plant herbaceous **117 Stackhousiaceae**

b Plant woody **41**

41a (40) Flowers (or at least some of them) functionally unisexual (i.e. anthers not producing pollen, ovary without ovules) **98 Simaroubaceae**

b Flowers functionally bisexual **42**

42a (41) Branches glaucous, ending in spines; leaves reduced to small, distant scales **115a Canotiaceae**

b Combination of characters not as above **43**

43a (42) Leaves entire or toothed; stamens 4–5, filaments free, emerging from the disc **115 Celastraceae**

b Leaves usually pinnate; stamens 8–10, filaments united into a tube, not emerging from the disc **100 Meliaceae**

44a (36) Plants herbaceous **45**

b Plants woody **49**

45a (44) Leaves always simple; ovary 6–10-celled by the development of 3–5 secondary septa during maturation **92 Linaceae**

b Leaves lobed or compound; secondary septa absent from the ovary **46**

46a (45) Leaves without stipules **47**

b Leaves with stipules **48**

47a (46) Ovary of 3–5 free carpels united only by a common style **87 Limnanthaceae**
b Ovary of 5 carpels whose bodies are completely united; styles 5, free **88 Oxalidaceae**
48a (46) Anthers 1-celled; leaves soft and mucilaginous; nectar secreted on the inner surfaces of the sepals **125 Malvaceae**
b Anthers 2-celled; leaves not soft and mucilaginous; nectar secreted round the base of the ovary **89 Geraniaceae**
49a (44) Filaments of the stamens united below **50**
b Filaments of stamens completely free from each other **52**
50a (49) Plants succulent, spiny; stamens 8, with woolly filaments; plants unisexual **25 Didieriaceae**
b Combination of characters not as above **51**
51a (50) Stipules persistent, borne between the bases of the leaf-stalks and the stems; petals appendaged **93 Erythroxylaceae**
b Stipules deciduous, not borne as above; petals not appendaged **127 Sterculiaceae**
52a (49) Stamens 8–10 **53**
b Stamens 2–6 **55**
53a (52) Petals long-clawed, often fringed or toothed; stamens 10; usually some or all of the sepals with nectar-secreting appendages on the outside **101 Malpighiaceae**
b Petals neither clawed nor toothed; stamens 8; sepals without nectar- secreting appendages **54**
54a (53) Leaves pinnate, exstipulate **107 Sapindaceae**
b Leaves simple, toothed, stipulate but stipules soon falling **132 Stachyuraceae**
55a (52) Stamens 2 **179 Oleaceae**
b Stamens 3–6 **56**
56a (55) Staminodes present in flowers that also contain fertile stamens **112 Cyrillaceae**
b Staminodes absent from flowers that also contain fertile stamens **57**

57a (56) Sepals united to each other at the base 58
b Sepals entirely free from each other 59
58a (57) Carpels 3, 1 or 2 of them sterile, the fertile containing 2 apical ovules **119 Icacinaceae**
b Carpels 3 or more, all fertile, each containing 1 or 2 apical ovules **113 Aquifoliaceae**
59a (57) Ovule 1 per cell; petals 3–4 **97 Cneoraceae**
b Ovules many per cell; petals 5 **78 Pittosporaceae**

Group VII

Dicotyledons with calyx and corolla present; petals free from each other; ovary of united carpels; stamens up to twice as many as petals; placentation axile, basal, apical or free-central; leaves opposite.

1a Petals and stamens numerous; plants succulent **19 Aizoaceae**
b Combination of characters not as above 2
2a (1) Placentation free-central, ovary of a single cell, at least above 3
b Placentation axile, basal or apical, ovary of 1 to several cells 4
3a (2) Sepals usually 2; if more, then petals numerous **20 Portulacaceae**
b Sepals or calyx-lobes 4 or 5; petals 4 or 5 **22 Caryophyllaceae**
4a (2) Corolla zygomorphic 5
b Corolla actinomorphic 7
5a (4) Plants woody; leaves palmate-digitate **108 Hippocastanaceae**
b Plants herbaceous; leaves various, not palmate-digitate 6
6a (5) Sepals, petals and stamens hypogynous **89 Geraniaceae**
b Sepals, petals and stamens perigynous **145 Lythraceae**
7a (4) Anthers opening by distinct terminal pores 8
b Anthers opening by longitudinal slits or valves 10

8a (7) Leaves with generally 3 parallel main veins from the base; each filament with a knee-like joint below the anther **150 Melastomataceae**
b Leaves with a single main vein from the base; filaments straight 9
9a (8) Stipules present; flowers borne in inflorescences, not solitary **123 Elaeocarpaceae**
b Stipules absent; flowers solitary **102 Tremandraceae**
10a (7) Small hairless annual herb growing in water or on wet mud; leaves with stipules; seeds pitted **139 Elatinaceae**
b Combination of characters not as above 11
11a (10) Sepals, petals and stamens perigynous 12
b Sepals, petals and stamens hypogynous 14
12a (11) Styles 2 or more; fruit an inflated, bladdery capsule; leaves trifoliolate or pinnate **116 Staphyleaceae**
b Style 1; fruit various, not as above; leaves simple 13
13a (12) Perigynous zone prominently ribbed; seeds without arils; mostly herbs **145 Lythraceae**
b Perigynous zone not ribbed; seeds with arils; shrubs or small trees **115 Celastraceae**
14a (11) Leaves with translucent, aromatic glands **96 Rutaceae**
b Leaves without such glands 15
15a (14) Flower with a well-developed disc, usually nectar-secreting, below and around the ovary 16
b Flower without a disc, nectar secreted in other ways 19
16a (15) Leaves often palmately lobed; sap sometimes milky; flowers functionally unisexual; fruit a group of winged samaras; trees **106 Aceraceae**
b Combination of characters not as above 17
17a (16) Leaves entire or toothed; stamens 4 or 5, emerging from the disc; seeds with arils **115 Celastraceae**
b Combination of characters not as above 18
18a (17) Leaves fleshy, with stipules; filaments free **91 Zygophyllaceae**

b Leaves not fleshy, exstipulate; filaments united into a tube **100 Meliaceae**

19a (15) Plant herbaceous **20**

b Plant woody **21**

20a (19) Leaves always simple and entire; ovary 6–10-celled by the development of 3–5 false septa during maturation; fruit a capsule **92 Linaceae**

b Leaves lobed or compound; ovary without false septa; fruit a schizocarp **89 Geraniaceae**

21a (19) Petals long-clawed, often fringed or toothed; stamens 10; usually some or all of the sepals with nectar-secreting appendages outside **101 Malpighiaceae**

b Petals not long-clawed, nor fringed or toothed; stamens 5; sepals without nectar-secreting appendages outside **78 Pittosporaceae**

Group VIII

Dicotyledons with sepals and petals; ovary superior; petals united into a tube at the base; corolla actinomorphic

1a Stamens 2, anthers back to back **179 Oleaceae**

b Stamens more than 2, anthers never back to back **2**

2a (1) Carpels several, free; leaves succulent **74 Crassulaceae**

b Carpels united, or, if the bodies of the carpels are free, then the styles united; leaves usually not succulent **3**

3a (2) Corolla papery, translucent, 4-lobed; stamens 4, projecting from the corolla; leaves with parallel veins, often all basal **210 Plantaginaceae**

b Combination of characters not as above **4**

4a (3) Central flowers of the inflorescence abortive, their bracts forming nectar-secreting pitchers; petals completely united, the corolla falling as a whole as the flower opens **61 Marcgraviaceae**

b Combination of characters not as above **5**

5a (4) Stamens more than twice as many as corolla-lobes 6
b Stamens up to twice as many as corolla-lobes 13
6a (5) Leaves evergreen, divided into 3 leaflets; filaments brightly coloured, at least twice as long as the petals **60 Caryocaraceae**
b Leaves deciduous or evergreen, simple, entire or lobed; filaments not as above 7
7a (6) Leaves with stipules; filaments of stamens united into a tube around the ovary and style **125 Malvaceae**
b Leaves without stipules; filaments free 8
8a (7) Anthers opening by pores **56 Actinidiaceae**
b Anthers opening by longitudinal slits 9
9a (8) Leaves with translucent, aromatic glands; calyx cup-like, unlobed **96 Rutaceae**
b Leaves without such glands; calyx not as above 10
10a (9) Placentation parietal; leaves fleshy **188 Fouquieriaceae**
b Placentation axile; leaves not fleshy 11
11a (10) Sap milky; ovules 1 per cell **175 Sapotaceae**
b Sap not milky; ovules 2 or more per cell 12
12a (11) Ovules 2 per cell; flowers usually unisexual **176 Ebenaceae**
b Ovules many per cell; flowers bisexual **59 Theaceae**
13a (5) Stamens as many as petals and on the same radii as them 14
b Stamens more or fewer than petals, if as many then not on the same radii as them 21
14a (13) Tropical trees with milky sap and evergreen leaves **175 Sapotaceae**
b Tropical or temperate trees, shrubs, herbs or climbers, with watery sap and usually deciduous leaves 15
15a (14) Placentation axile 16
b Placentation basal or free-central 17
16a (15) Climbers with tendrils; stamens free **121 Vitaceae**

b Upright shrubs without tendrils; stamens with the filaments united below **122 Leeaceae**

17a (15) Trees or shrubs; fruit a berry or drupe **18**

b Herbs (occasionally woody at the extreme base); fruit a capsule or indehiscent **19**

18a (17) Leaves with translucent glands; anthers opening towards the centre of the flower; staminodes absent **172 Myrsinaceae**

b Leaves without such glands; anthers opening towards the outside of the flower; staminodes 5 **171 Theophrastaceae**

19a (17) Sepals 2, free **20 Portulacaceae**

b Sepals 4 or more, united **20**

20a (19) Corolla persistent and papery in fruit; ovule 1 on a long stalk arising from the base of the ovary **174 Plumbaginaceae**

b Corolla not persistent and papery in fruit; ovules many, on a free-central placenta **173 Primulaceae**

21a (13) Flower compressed with 2 planes of symmetry; stamens united in 2 bundles of $\frac{1}{2} + 1 + \frac{1}{2}$ **66a Fumariaceae**

b Combination of characters not as above **22**

22a (21) Leaves bipinnate or replaced by phyllodes; carpel 1; fruit a legume **84b Mimosaceae**

b Combination of characters not as above **23**

23a (22) Anthers opening by pores (rarely by short, pore-like slits); pollen never in coherent masses **24**

b Anthers opening by longitudinal slits or pollen in coherent masses (pollinia) **25**

24a (23) Stamens free from corolla-tube, often twice as many as corolla-lobes **168 Ericaceae**

b Stamens borne on the corolla-tube, as many as lobes **197 Solanaceae**

25a (23) Leaves alternate or all basal; carpels never 2 and free or almost so but united by the common style **26**

b Leaves opposite or rarely alternate, when the carpels are 2 and almost completely free, united by the common style **46**

26a (25) Flowers unisexual; male flowers with a corolla, female flowers without a corolla **94 Euphorbiaceae**
b Flowers bisexual, all with corollas **27**
27a (26) Plant woody; leaves usually evergreen, often spiny-margined; stigma sessile, on top of the ovary **113 Aquifoliaceae**
b Combination of characters not as above **28**
28a (27) Shrubs with stellate hairs or lepidote scales **177 Styracaceae**
b Herbs or shrubs, without stellate hairs or lepidote scales **29**
29a (28) Procumbent herbs with milky sap and stamens free from the corolla-tube **215 Campanulaceae**
b Combination of characters not as above **30**
30a (29) Ovary 5-celled **31**
b Ovary 2–4-celled **33**
31a (30) Placentation parietal; soft-wooded tree **140 Caricaceae**
b Placentation axile; herbs **32**
32a (31) Leaves fleshy; anthers 2-celled; fruit often deeply lobed **196 Nolanaceae**
b Leaves leathery; anthers 1-celled; fruit a capsule or berry **170 Epacridaceae**
33a (30) Ovary 3-celled **34**
b Ovary 1-, 2- or 4-celled **37**
34a (33) Trees; stamens free from the corolla-tube **13 Olacaceae**
b Shrubs, herbs or climbers; stamens borne on the corolla-tube **35**
35a (34) Dwarf, evergreen shrublets; staminodes 5; petals imbricate **165 Diapensiaceae**
b Herbs or climbers, not evergreen; staminodes absent; petals contorted **36**
36a (35) Climber with tendrils **187a Cobaeaceae**
b Herbs, without tendrils **187 Polemoniaceae**
37a (33) Stamens with filaments united into a tube; flowers in heads; stigmas surrounded by a sheath **217 Brunoniaceae**

b Combination of characters not as above 38

38a (37) Flowers in spirally coiled cymes, or the calyx with appendages between the lobes; style terminal or arising from between the lobes of the ovary 39

b Flowers not in spirally coiled cymes, calyx without appendages; style terminal 40

39a (38) Style terminal; fruit a capsule, usually many-seeded **190 Hydrophyllaceae**

b Style arising from the depression between the 4 lobes of the ovary; fruit of 4 nutlets, or more rarely a 1–4-seeded drupe **191 Boraginaceae**

40a (38) Placentation parietal 41

b Placentation axile 42

41a (40) Corolla-lobes valvate in bud; leaves simple and cordate or peltate, or of 3 leaflets, hairless; aquatic or marsh plants **183 Menyanthaceae**

b Corolla lobes imbricate in bud; leaves never as above; terrestrial plants **205 Gesneriaceae**

42a (40) Ovules 1–2 in each cell of the ovary 43

b Ovules 3 – many in each cell of the ovary 45

43a (42) Arching shrubs with small purple flowers in clusters on the previous year's wood **198 Buddlejaceae**

b Combination of characters not as above 44

44a (43) Sepals free; corolla-lobes contorted and infolded in bud; twiners, herbs or dwarf shrubs **189 Convolvulaceae**

b Sepals united; corolla-lobes not as above in bud; trees or shrubs **191 Boraginaceae**

45a (42) Corolla-lobes folded, valvate or contorted in bud; septum of the ovary oblique, not in the horizontal plane **197 Solanaceae**

b Corolla-lobes variously imbricate but not as above in bud; septum of ovary in the horizontal plane **199 Scrophulariaceae**

46a (25) Trailing, heather-like shrublet **168 Ericaceae**
b Plant not as above **47**
47a (46) Milky sap usually present; fruit usually of 2 almost free follicles united by a common style; seeds with silky appendages **48**
b Milky sap absent; fruit a capsule or fleshy, carpels united; seeds without silky appendages **49**
48a (47) Pollen granular; corona absent; style with a swelling below the stigma **184 Apocynaceae**
b Pollen usually in coherent masses (pollinia); corona usually present; style without a swelling below the stigma **185 Asclepiadaceae**
49a (47) Root-parasites without chlorophyll **192 Lennoaceae**
b Free-living plants with chlorophyll **50**
50a (48) Flowers in coiled cymes; usually herbs **190 Hydrophyllaceae**
b Flowers not in coiled cymes; herbs or shrubs **51**
51a (50) Placentation parietal; carpels 2 **52**
b Placentation axile; carpels 2, 3 or 5 **53**
52a (51) Leaves compound; epicalyx present **190 Hydrophyllaceae**
b Leaves simple; epicalyx absent **182 Gentianaceae**
53a (51) Stamens fewer than corolla-lobes **193 Verbenaceae**
b Stamens as many as corolla-lobes **54**
54a (53) Carpels 5; shrubs with leaves with spiny margins **181 Desfontainiaceae**
b Carpels 2 or 3; herbs or shrubs; leaves not as above **55**
55a (54) Leaves without stipules; carpels 3; corolla-lobes contorted in bud; herbs **187 Polemoniaceae**
b Leaves with stipules (often reduced to a ridge between the leaf-bases); corolla-lobes variously imbricate or valvate in bud; plant usually woody **56**

56a (55) Corolla usually 5-lobed; stellate and/or glandular hairs absent **180 Loganiaceae**
b Corolla 4-lobed; stellate and glandular hairs present **198 Buddlejaceae**

Group IX

Dicotyledons with sepals and petals; ovary superior; petals united into a tube at the base; corolla zygomorphic

1a Stamens more numerous than the corolla-lobes, or anthers opening by pores 2
b Stamens as many as corolla-lobes or fewer; anthers not opening by pores 6
2a (1) Anthers opening by pores; leaves undivided; ovary of 2 or more united carpels 3
b Anthers opening by longitudinal slits; leaves dissected or compound; ovary of a single carpel 5
3a (2) The 2 lateral sepals large and petal-like; filaments united **103 Polygalaceae**
b No sepals petal-like; filaments free 4
4a (3) Shrubs with alternate or apparently whorled leaves; stamens 4–27 **168 Ericaceae**
b Herbs with opposite leaves; stamens 5 **182 Gentianaceae**
5a (2) Leaves pinnate or of 3 leaflets; perianth not spurred **84 Fabaceae**
b Leaves laciniate; upper petal spurred; upper sepal helmet-like or spurred **41 Ranunculaceae**
6a (1) Stamens as many as corolla-lobes; zygomorphy of corolla usually weak 7
b Stamens fewer than corolla-lobes; zygomorphy of corolla pronounced 14
7a (6) Stamens on the same radii as the corolla-lobes; placentation free-central **173 Primulaceae**

b Stamens on different radii from the corolla-lobes; placentation axile 8

8a (7) Leaves of 3 leaflets, with translucent, aromatic glands; stamens 5, the upper 2 fertile, the lower 3 sterile **96 Rutaceae**

b Combination of characters not as above 9

9a (8) Ovary of 3 carpels; ovules many **187 Polemoniaceae**

b Ovary of 2 carpels; ovules 4 or many 10

10a (9) Flowers in coiled cymes; fruit of up to 4 1-seeded nutlets **191 Boraginaceae**

b Flowers not in coiled cymes; fruit a many-seeded capsule 11

11a (10) Annual or shortly-lived perennial climber; corolla scarlet at first, fading to yellow-white **189 Convolvulaceae**

b Combination of characters not as above 12

12a (11) Corolla-lobes variously imbricate in bud; stamens 2, 4 or 5 and unequal; leaves usually alternate **199 Scrophulariaceae**

b Corolla-lobes contorted in bud; stamens 5, equal 13

13a Leaves opposite; woody climber **180 Loganiaceae**

b Leaves alternate; annual or perennial herbs **197 Solanaceae**

14a (6) Placentation axile; ovules 4 or many 15

b Placentation parietal, free-central, apical or basal; ovules many or 1 or 2 22

15a (14) Ovules numerous but not in vertical rows in each cell of the ovary 16

b Ovules 4, or more numerous but then in vertical rows in each cell of the ovary 18

16a (15) Seeds winged; mainly trees, shrubs and climbers with opposite, pinnate, digitate or rarely simple leaves **201 Bignoniaceae**

b Seeds usually wingless; mainly herbs or shrubs with simple leaves 17

17a (16) Corolla-lobes imbricate in bud; septum of the ovary in the horizontal plane; leaves opposite or alternate **199 Scrophulariaceae** b Corolla-lobes usually folded, contorted or valvate in bud; septum of ovary oblique, not in the horizontal plane; leaves alternate **197 Solanaceae**

18a (15) Leaves all alternate, usually with blackish, resinous glands; plants woody **208 Myoporaceae**

b At least the lower leaves opposite or whorled, none with glands as above; plants herbaceous or woody **19**

19a (18) Fruit a capsule; ovules 4 to many, usually in vertical rows in each cell of the ovary **20**

b Fruit not a capsule; ovules 4, side-by-side **21**

20a (19) Leaves all opposite, often prominently marked with cystoliths; flower-stalks without swollen glands at the base; capsule usually opening elastically, seeds usually on hooked stalks **202 Acanthaceae**

b Upper leaves alternate, cystoliths absent; flower-stalks with swollen glands at the base; capsule not elastic, seeds not on hooked stalks **203 Pedaliaceae**

21a (19) Style arising from the depression between the 4 lobes of the ovary, or if terminal then corolla with a reduced upper lip; fruit usually of 4 1-seeded nutlets; calyx and corolla often 2-lipped **195 Labiatae**

b Style terminal; corolla with well-developed upper lip; fruit usually a berry or drupe; calyx often more or less actinomorphic, not 2-lipped **193 Verbenaceae**

22a (14) Ovules 4 to many; fruit a capsule, rarely a berry or drupe **23**

b Ovules 1–2; fruit indehiscent, often dispersed in the persistent calyx **29**

23a (22) Ovary containing 4 ovules side-by-side **193 Verbenaceae**

b Ovary containing many ovules **24**

24a (23) Placentation free-central; corolla spurred; leaves modified for trapping and digesting insects **207 Lentibulariaceae**

b Placentation parietal or apical; corolla not spurred, rarely swollen at base; leaves not insectivorous **25**
25a (24) Leaves scale-like, never green; root-parasites **26**
b Leaves green, expanded; free-living plants **27**
26a (25) Placentas 4; calyx laterally 2-lipped **206 Orobanchaceae**
b Placentas 2; calyx 4-lobed **199 Scrophulariaceae**
27a (25) Seeds winged; mainly climbers with opposite, pinnately divided leaves **201 Bignoniaceae**
b Combination of characters not as above **28**
28a (27) Capsule with a long beak separating into 2 curved horns; plants sticky-velvety **204 Martyniaceae**
b Capsule without beak or horns; plant velvety or variously hairy or hairless **205 Gesneriaceae**
29a (22) Flowers in heads surrounded by an involucre of bracts; ovule 1 **200 Globulariaceae**
b Flowers not in heads, often in spikes; ovules 1 or 2 **30**
30a (29) Fruits deflexed; calyx with hooked teeth; ovary 1-celled with 1 basal ovule **209 Phrymaceae**
b Fruits mostly erect; calyx without hooked teeth; ovary 2-celled with a solitary apical ovule in each cell; fruit often 1-seeded **199 Scrophulariaceae**

Group X

Dicotyledons with perianth of a single whorl, not distinguished into sepals and petals; at least the male flowers borne in catkins, which are usually deciduous as a whole.

1a Stems jointed; leaves reduced to whorls of scales **1 Casuarinaceae**
b Stems not jointed; leaves not as above **2**
2a (1) Leaves pinnate **3**

b Leaves simple and entire, toothed or lobed (sometimes deeply so) 4

3a (2) Leaves without stipules; fruit a nut **3 Juglandaceae**

b Leaves with stipules; fruit a legume **84a Caesalpiniaceae**

4a (2) Male flowers with a perianth of 2 segments and 4 or 5 fertile stamens plus 4 or 5 staminodes; female flowers without a perianth; shrubs; fruit a syncarp of berries **71 Bataceae**

b Combination of characters not as above 5

5a (4) Leaves opposite, evergreen, entire; fruit berry-like **162 Garryaceae**

b Leaves alternate, deciduous or evergreen; fruit not berry-like 6

6a (5) Ovules many, parietal; seeds many, cottony-hairy; male catkin erect with the stamens projecting between the bracts, or hanging and with fringed bracts **5 Salicaceae**

b Ovules solitary or few, not parietal; seeds few, not cottony-hairy; male catkins not as above 7

7a (6) Leaves dotted with aromatic glands **2 Myricaceae**

b Leaves not dotted with aromatic glands 8

8a (7) Styles 3, each often branched; fruit splitting into 3 mericarps; seeds with appendages **94 Euphorbiaceae**

b Styles 1–6, not branched; fruit and seeds not as above 9

9a (8) Plant with milky sap **10 Moraceae**

b Plant with clear sap 10

10a (9) Male catkin compound, i.e. each bract with 2–3 flowers attached to it; styles 2 11

b Male catkin simple, i.e. each bract with a single flower attached to it; styles 1 or 3–6 12

11a (10) Nuts small, borne in cone-like catkins; perianth present in male flowers, absent in female, ovary naked **6 Betulaceae**

b Nuts large, subtended by leaf-like bracts or involucres (cupules); perianth present in female flowers, absent in male; ovary inferior **6a Corylaceae**

12a (10) Ovary inferior; fruit a nut surrounded or enclosed by a scaly cupule; stipules deciduous; styles 3–6 **7 Fagaceae** b Ovary superior; fruit a leathery drupe, cupule absent; stipules absent; style 1 **4 Leitneriaceae**

Group XI

Dicotyledons with perianth of a single whorl, not distinguished into sepals and petals; catkins absent; ovary of a single carpel or made up of several free carpels.

1a Ovary apparently of a single carpel **2**
b Ovary of 2 or more free carpels **17**
2a (1) Mostly submerged aquatic herbs with at least the submerged leaves whorled **3**
b Terrestrial plants, sometimes growing in damp places; leaves not whorled **4**
3a (2) Leaves much divided; stamens 10–20, borne beneath the ovary **46 Ceratophyllaceae**
b Leaves simple, entire; stamen 1, borne on the upper part of the ovary **156 Hippuridaceae**
4a (2) Root-parasites without chlorophyll; leaves scale-like **157 Cynomoriaceae**
b Free-living plants with chlorophyll, leaves well developed **5**
5a (4) Leaves with stipules **6**
b Leaves without stipules **10**
6a (5) Rhubarb-like marsh plants with large leaves; stamens 1 or 2 **154a Gunneraceae**
b Combination of characters not as above **7**
7a (6) Herbs or soft-wooded shrubs, often with stinging hairs; cystoliths present in the leaves; stamens 4 or 5, inflexed in bud, exploding when ripe **11 Urticaceae**
b Combination of characters not as above **8**
8a (7) Leaves opposite; flowers unisexual **49 Chloranthaceae**
b Leaves alternate; flowers bisexual **9**
9a (8) Stamens 4; epicalyx present **82 Rosaceae**

b Stamens 5–7; epicalyx absent **83 Chrysobalanaceae**
10a (5) Herbs; leaves with sheathing bases, the lower opposite, the upper alternate **155 Theligonaceae**
b Shrubs or trees; leaves not as above 11
11a (10) Stamens borne on the perianth 12
b Stamens free from the perianth 14
12a (11) Trees or shrubs with very hard, leathery leaves; perianth segments free, usually spoon-shaped **12 Proteaceae**
b Shrubs; leaves deciduous or evergreen but not very hard; perianth-segments united into a tube below 13
13a (12) Plants covered in lepidote scales; ovule basal **129 Elaeagnaceae**
b Plants not covered in lepidote scales; ovule apical **128 Thymelaeaceae**
14a (11) Large evergreen trees or shrubs 15
b Herbs or small, deciduous shrubs 16
15a (14) Plants aromatic; leaves glandular-punctate; anthers opening by valves **36 Lauraceae**
b Plants not aromatic; leaves not glandular-punctate; anthers opening by longitudinal slits **30 Myristicaceae**
16a (14) Flowers in racemes; fruit often fleshy; stamens 3 – many **17 Phytolaccaceae**
b Flowers in cymes; fruit an achene; stamens usually 5 **18 Nyctaginaceae**
17a (1) Trees with bark peeling off in plates; leaves palmately lobed, base of petiole covering the axillary bud; flowers unisexual in hanging, spherical heads **72 Platanaceae**
b Combination of characters not as above 18
18a (17) Perianth completely absent 19
b Perianth present 20
19a (18) Herbs **47 Saururaceae**
b Small trees or shrubs **39 Eupteleaceae**
20a (18) Perianth and stamens perigynous, borne on a rim or cup, itself borne below the ovary 21

b Perianth and stamens hypogynous, borne independently below the ovary 24

21a (20) Leaves modified into insect-trapping pitchers **75 Cephalotaceae**

b Leaves not modified into pitchers 22

22a (21) Flowers unisexual; leaves evergreen **34 Monimiaceae**

b Flowers bisexual; leaves deciduous 23

23a (22) Inner stamens sterile; perianth of many segments; leaves opposite **35 Calycanthaceae**

b Stamens all fertile; perianth of up to 9 segments; leaves usually alternate **82 Rosaceae**

24a (20) Leaves with conspicuous stipules which enclose the axillary buds; bark aromatic **27 Magnoliaceae**

b Leaves without stipules; bark usually not aromatic 25

25a (24) Woody climbers 26

b Herbs, shrubs or trees 30

26a (25) Leaves opposite; flowers bisexual; plant climbing by means of hooked, hardened petioles **41 Ranunculaceae**

b Leaves alternate; flowers unisexual; plant twining 27

27a (26) Leaves compound; parts of the flower in 3s **43 Lardizabalaceae**

b Leaves simple; parts of the flower not usually in 3s 28

28a (27) Leaves evergreen, leathery, wavy-margined; flowers in dense, cone-like racemes; carpels 5 or more, each 1-seeded **17 Phytolaccaceae**

b Combination of characters not as above 29

29a (27) Carpels many; seeds not U-shaped **32 Schisandraceae**

b Carpels 3 or 6; seeds usually U-shaped **44 Menispermaceae**

30a (25) Parts of the flower in 3s; fruits blue **43 Lardizabalaceae**

b Combination of characters not as above 31

31a (30) Perianth-segments 6 or more in 2–3 whorls, sometimes differing a little in size and colour; bark aromatic **33 Illiciaceae** b Combination of characters not as above **32**

32a (31) Trees with rounded, cordate leaves, which are opposite on long shoots, alternate on short shoots; flowers axillary, very inconspicuous, unisexual **40 Cercidiphyllaceae**

b Combination of characters not as above **33**

33a (32) Each anther tipped by an enlarged connective; fruit a berry or an aggregate of berries; plants woody **29 Annonaceae**

b Anthers not tipped by enlarged connectives; fruit not as above; plants usually herbaceous **34**

34a (33) Stamens 5 or rarely more; prostrate herb with opposite or almost whorled leaves **17d Gisekiaceae**

b Stamens numerous; habit and leaves not as above **41 Ranunculaceae**

Group XII

Dicotyledons with perianth of a single whorl, not distinguished into sepals and petals; catkins absent; ovary of several united carpels; stamens borne on the perianth or ovary inferior.

1a Plants aquatic, mostly submerged 2

b Plants terrestrial 4

2a (1) Stamens 8, 4 or 2; leaves deeply divided **154 Haloragaceae**

b Stamens 6 or 1; leaves entire or slightly toothed 3

3a (2) Stamens 6; leaves all basal **68 Cruciferae**

b Stamen 1; leaves opposite **194 Callitrichaceae**

4a (1) Trees or shrubs 5

b Herbs, climbers or parasites 13

5a (4) Stamens as many as, and on radii alternating with, the perianth-segments **120 Rhamnaceae**

b Stamens not as above 6

6a (5) Stipules present, sometimes falling early 7

b Stipules absent 9

7a (6) Styles 3–6; fruit a nut, surrounded by a scaly cupule **7 Fagaceae**
b Styles 2; fruit not as above 8
8a (7) Leaves alternate; stellate hairs usually present; fruit a woody capsule **73 Hamamelidaceae**
b Leaves opposite; stellate hairs absent; fruit a non-woody capsule **77 Cunoniaceae**
9a (6) Ovary superior; leaves opposite; sap sometimes milky; fruit a group of samaras; trees **106 Aceraceae**
b Ovary inferior; combination of other characters not as above 10
10a (9) Ovary 1-celled; ovule 1, apical, or ovules 1–5, basal 11
b Ovary several-celled, or if 1-celled then ovules more than 5, parietal or axile 12
11a (10) Epigynous zone present above the ovary, bearing the perianth on its rim and stamens on its inner face; ovule 1, apical **152 Combretaceae**
b Epigynous zone absent, perianth and stamens not as above; ovules 1–5, basal **14 Santalaceae**
12a (10) Placentation parietal; flowers bisexual, variously arranged but not as below **76c Hydrangeaceae**
b Placentation axile; flowers unisexual in heads consisting of many male flowers surrounding a single female flower, each head subtended by 2 large, white bracts **160 Davidiaceae**
13a (4) Plants parasitic 14
b Plants free-living 18
14a (13) Branch-parasites with green, forked branches or stalkless flowers borne directly on the branches of the host 15
b Root-parasites, lacking chlorophyll 17
15a (14) Flowers (the only visible parts of the plant) brown, minute, sessile on the branches of the host **51 Rafflesiaceae**
b. Flowers borne on green, forked branches 16
16a (15) Two united bracteoles forming a cup-like structure, borne just below the perianth **15 Loranthaceae**

b Bracteoles absent **15a Viscaceae**

17a (14) Flowers minute, in fleshy spikes; stamen 1 **157 Cynomoriaceae**

b Flowers conspicuous in short, bracteate spikes; stamens more than 1 **51 Rafflesiaceae**

18a (13) Perianth absent; flowers in spikes **47 Saururaceae**

b Perianth present; flowers not usually in spikes 19

19a (18) Leaf-base oblique; ovary inferior, 3-celled **143 Begoniaceae**

b Leaf-base not oblique; ovary not as above 20

20a (19) Ovary superior 21

b Ovary inferior 27

21a (20) Carpels 3 or rarely 2, ovule 1, basal; perianth persistent in fruit 22

b Combination of characters not as above 23

22a (21) Leaves without stipules; stamens 5 **21 Basellaceae**

b Leaves with stipules usually united into a sheath (ochrea) around the stem; stamens usually 6–9 **16 Polygonaceae**

23a (21) Annual herbs with leaves in apparent whorls; placentation axile; perianth and stamens hypogynous **17c Molluginaceae**

b Combination of characters not as above 24

24a Carpels 5 or rarely 6, united only below, free above **76a Penthoraceae**

b Carpels either 1 or 2–3, fully united 25

25 (24) Leaves alternate, usually lobed or compound **82 Rosaceae**

b Leaves usually opposite, entire 26

26a (25) Ovule 1, basal; fruit a nut; stipules usually present, often hyaline **22a Illecebraceae**

b Ovules numerous; fruit a capsule; stipules absent **145 Lythraceae**

27a (20) Leaves pinnate; ovary open at the apex **142 Datiscaceae**

b Leaves not pinnate; ovary closed at apex 28

28a (27) Ovary 6-celled; perianth 3-lobed, or tubular and bilaterally symmetric **50 Aristolochiaceae**
b Combination of characters not as above **29**
29a (28) Ovules 1–5; seed 1 **30**
b Ovules and seeds numerous **31**
30a (29) Perianth-segments thickening in fruit; leaves alternate **23 Chenopodiaceae**
b Perianth-segments not thickening in fruit; leaves opposite or alternate **14 Santalaceae**
31a (29) Styles 2; placentation parietal **76c Hydrangeaceae**
b Style 1; placentation axile **32**
32a (31) Stamens 15–30 **17a Agdestidaceae**
b Stamens 8 or fewer **153 Onagraceae**

Group XIII

Dicotyledons with perianth of a single whorl, not distinguished into sepals and petals; catkins absent; ovary of several united carpels; stamens free from the perianth or ovary superior.

1a Aquatic plants, either submerged in part or at least partly covered by flowing water **2**
b Terrestrial plants **3**
2a (1) Plants not clearly differentiated into stem and leaves, thallus-like; perianth zygomorphic; stamens 1–4 **86 Podostemaceae**
b Plants with well-differentiated stems and leaves; perianth actinomorphic; stamen 1 **194 Callitrichaceae**
3a (1) Climbers or scramblers, most leaves ending in a tendril-like structure, which itself terminates in an insectivorous pitcher **64 Nepenthaceae**
b Combination of characters not as above **4**
4a (3) Stipules present, sometimes falling early **5**
b Stipules entirely absent **17**

5a (4) Ovary 1-celled, containing a single ovule 6
b Ovary 1- to several-celled, containing 2 or more ovules 10
6a (5) Styles 2–4, usually 3, free; stipules sheathing the stems **16 Polygonaceae**
b Style 1, sometimes deeply divided above into 2 stigmas; stipules not as above 7
7a (6) Ovule basal; herbs or shrubs, flowers never sunk in a fleshy receptacle, leaves never palmately lobed or divided; cystoliths present **11 Urticaceae**
b Ovule apical; trees, shrubs, herbs or climbers, if herbs then flowers sunk in a fleshy receptacle or leaves palmately lobed or divided; cystoliths absent 8
8a (7) Herbs or climbers; perianth in male flowers of 5 united segments **10a Cannabaceae**
b Trees or shrubs; perianth not as above 9
9a (8) Sap watery; stigma 1; flowers often bisexual; leaves usually oblique at the base **8 Ulmaceae**
b Sap milky; stigmas 2; flowers usually unisexual; leaves not oblique at the base **10 Moraceae**
10a (5) Placentation parietal or free-central 11
b Placentation axile, apical or basal 12
11a (10) Shrubs, trees or climbers; leaves alternate; placentation parietal **130 Flacourtiaceae**
b Herbs; leaves usually opposite; placentation free-central **22 Caryophyllaceae**
12a (10) Leaves large, pinnate; stipules large, palmately veined; irritant hairs present **77a Davidsoniaceae**
b Combination of characters not as above 13
13a (12) Sap milky; styles usually 3, often divided; ovules 1–2 per cell **94 Euphorbiaceae**
b Combination of characters not as above 14
14a (13) Stellate hairs usually present 15
b Stellate hairs absent 16

15a (14) Flowers unisexual; stamens with free filaments; carpels 2, styles diverging **73 Hamamelidaceae**
b Flowers bisexual; stamens with their filaments united into a tube around the styles; ovary not as above **127 Sterculiaceae**
16a (14) Style 1; trees or shrubs **123 Elaeocarpaceae**
b Styles 3–4; herbs **47 Saururaceae**
17a (4) Trees with milky sap; styles 2 **9 Eucommiaceae**
b Combination of characters not as above **18**
18a (17) Ovary 1-celled, containing a single basal ovule **19**
b Ovary 1–several-celled, containing several ovules **25**
19a (18) Flowers minute, bisexual, usually sunk in a fleshy spike; stigma sessile, usually brush-like **48 Piperaceae**
b Combination of characters not as above **20**
20a (19) Leaves usually with stipules united into a sheath (ochrea) around the stem; young leaves revolute; stamens 6–9; styles 2–4, usually 3, free **16 Polygonaceae**
b Combination of characters not as above **21**
21a (20) Leaves usually opposite or whorled; fruit an achene borne in the persistent perianth; style 1, slightly lobed at apex **18 Nyctaginaceae**
b Combination of characters not as above **22**
22a (21) Lower leaves opposite, upper alternate; perianth tubular, bulging conspicuously at one side in the female flower **155 Theligonaceae**
b Leaves all alternate or all opposite; perianth not as above **23**
23a (22) Stamens 10–2-; fruit a translucent berry **17b Achatocarpaceae**
b Stamens up to 5, occasionally with some additional staminodes; fruit not a translucent berry **24**
24a (23) Filaments united below; perianth usually hyaline and/or papery **24 Amaranthaceae**
b Filaments free; perianth herbaceous **23 Chenopodiaceae**
25a (18) Plants woody **26**

b Plants herbaceous **36**

26a (25) Creeping shrublets with heather-like leaves **169 Empetraceae**

b Trees or upright shrubs; leaves not heather-like **27**

27a (26) Leaves opposite or whorled **28**

b Leaves alternate **31**

28a (27) Leaves whorled; carpels 6–10, each with an individual stigma **38 Trochodendraceae**

b Leaves opposite; carpels 2 or 3, style usually 1 **29**

29a (28) Stamens 2, rarely 3 or 1; leaves usually deciduous, often pinnate **179 Oleaceae**

b Stamens 4–12; leaves evergreen, simple **30**

30a (29) Stamens 4 **118 Buxaceae**

b Stamens 8–12 **118a Simmondsiaceae**

31a (27) Leaves compound or deeply divided **107 Sapindaceae**

b Leaves simple, entire or slightly lobed or toothed **32**

32a (31) Ovary 2-celled below, 1-celled above; leaves evergreen; flowers unisexual, males with a perianth of 3–8 imbricate segments, females without a perianth **95 Daphniphyllaceae**

b Combination of characters not as above **33**

33a (32) Resinous trees and shrubs; hypogynous disc usually present; ovary 3–5-celled **105 Anacardiaceae**

b Non-resinous trees and shrubs; hypogynous disc absent; ovary usually 6- or more-celled **34**

34a (33) Connective of stamens prolonged as appendages; fruit an aggregate of berries **29 Annonaceae**

b Connective of stamens not appendaged; fruit not as above **35**

35a (34) Perianth-segments, stamens and carpels 4 each; inflorescence catkin-like **37 Tetracentraceae**

b Perianth-segments, stamens and carpels each more numerous; inflorescence a raceme **17 Phytolaccaceae**

36a (25) Ovary 1-celled, placentation free-central; stamens as many as and on alternating radii with the perianth-segments **173 Primulaceae** b Ovary 1- or more-celled, placentation not free-central; stamens not as above **37**

37a (36) Leaves modified into insect-trapping pitchers; style peltate **63 Sarraceniaceae**

b Leaves not modified into insect-trapping pitchers; style not peltate **38**

38a (37) Placentation parietal; perianth-segments 2, falling quickly **66 Papaveraceae**

b Placentation basal or axile; perianth-segments more than 2, not falling quickly **39**

39a (38) Perianth-segments 4, mauve; stamens very numerous, bright yellow; carpels 2, united only at the base, and diverging **41a Glaucidiaceae**

b Perianth and stamens not as above; carpels several, united for most of their length, not diverging 40

40a (39) Placentation axile, ovules several per cell **17c Molluginaceae**

b Placentation basal; ovules 1 per cell **17 Phytolaccaceae**

Group XIV

Monocotyledons with superior ovaries, or plants aquatic with totally submerged flowers.

1a Trees, shrubs or prickly scramblers with large, pleated, usually palmately or pinnately divided leaves; flowers more or less stalkless in fleshy spikes or panicles which often have large, sometimes woody, basal bracts (spathes) **247 Palmae**

b Combination of characters not as above 2

2a (1) Totally submerged aquatic plants of fresh or saline water 3

b Terrestrial or epiphytic plants, if aquatic then not totally submerged, occasionally entirely floating 10

3a (2) Perianth of 4 clawed, valvate segments; freshwater aquatics with bisexual flowers in submerged or emergent spikes; carpels 4, free **226 Potamogetonaceae**

b Combination of characters not as above 4
4a (3) Plants growing in the sea or in strongly brackish water 5
b Plants of fresh water 8
5a (4) Flowers bisexual; style irregularly branched or dilated 6
b Flowers unisexual; style not irregularly branched or dilated 7
6a (5) Marine plants with densely fibrous rhizomes; leaves basal, with ligules **226c Posidoniaceae**
b Plants of brackish water; leaves opposite, without ligules **226a Ruppiaceae**
7a (5) Flowers borne on one side of a flattened spike **226b Zosteraceae**
b Flowers in cymose clusters or solitary in the leaf-axils **227a Cymodoceaceae**
8a (4) Ovules 2 in each of the 3 or 6 free carpels; fruit a group of follicles **223 Scheuchzeriaceae**
b Ovules 1 in each of the 1–9 free carpels, or 1 in the 1-celled ovary, which has 2–4 stigmas; fruits indehiscent 9
9a (8) Ovary 1-celled with 2–4 styles, ovule 1, basal **228 Najadaceae**
b Ovary of several carpels, each with a single ovule and style **227 Zannichelliaceae**
10a (2) Small floating aquatic plants not differentiated into stem and leaves **249 Lemnaceae**
b Plants aquatic or not, clearly differentiated into stem and leaves (the latter rarely scale-like) 11
11a (10) Perianth hyaline or papery, translucent or brown, or reduced to bristles, hairs, narrow scales or totally absent 12
b Perianth well-developed, though sometimes small, not hyaline or papery, usually coloured, sometimes greenish 20
12a (11) Flowers in small, 2-sided or cylindric spikelets provided with overlapping bracts (spikelets sometimes 1-flowered) 13
b Flowers arranged in heads, superposed spikes, racemes, panicles or cymes, never in spikelets as above 14

13a (12) Leaves alternate in 2 ranks on a stem, which is usually hollow and with cylindric internodes; leaf-sheath usually with free margins, at least in the upper part; flowers arranged in 2-sided spikelets (sometimes 1-flowered) each usually subtended by 2 sterile bracts (glumes); each flower usually enclosed by bracts (a lower lemma and an upper palea (sometimes absent)); perianth of 2 or 3 concealed scales (lodicules), more rarely of 6 scales or absent; styles generally 2, feathery **246 Gramineae**

b Leaves usually arranged on 3 sides of the cylindric or more usually 3-angled stems, which usually have solid internodes; young leaf-sheaths closed, though sometimes splitting later; flowers arranged in 2-sided or cylindric spikelets, often with a 2-keeled or 2-lobed glume at the base; each flower subtended by a glume only; perianth of several bristles, hairs, scales or absent; style 1 with 2 or 3 papillose stigmas **253 Cyperaceae**

14a (12) Dioecious trees or shrubs often supported by stilt-roots, with stiffly leathery, sharply toothed leaves; fruits compound, often woody **250 Pandanaceae**

b Combination of characters not as above **15**

15a (14) Inflorescence a simple, fleshy spike (spadix) of inconspicuous flowers, subtended by or rarely joined to a large bract which encloses it in bud (spathe); leaves often net-veined and/or lobed (plant rarely a small evergreen floating aquatic) **16**

b Inflorescence various but not as above; leaves not as above **17**

16a (15) Leaves iris-like, parallel-veined, fragrant when crushed; spadix protruding obliquely sideways from what appears to be a normal foliage leaf **248a Acoraceae**

b Leaves various, but not as above, usually net-veined and/or lobed; spike and spadix various, not as above **248 Araceae**

17a (15) Flowers unisexual, in heads each of which is surrounded by an involucre of bracts; perianth segments in 2 series, often greyish white **245 Eriocaulaceae**

b Combination of characters not as above 18

18a (17) Flowers bisexual; perianth-segments 6, hyaline or brownish; ovary with 3 to many ovules **240 Juncaceae**

b Flowers unisexual; perianth-segments a few threads or scales; ovary with 1 ovule 19

19a (18) Flowers in 2 superposed, elongate, brownish or silvery spikes; ovary borne on a stalk, which has hair-like branches **252 Typhaceae**

b Flowers in spherical heads; ovary not stalked **251 Sparganiaceae**

20a (11) Carpels free, or very slightly united at the base only, or ovary apparently of a single carpel with a single style 21

b Carpels united for most of their length, though the styles may be free 27

21a (20) Female flowers of 2 kinds, solitary in the leaf-axils and very long-styled and several in bracteate spikes, with short styles **225a Lilaeaceae**

b Combination of characters not as above 22

22a (21) Inflorescence a spike, sometimes bifid; perianth-segments 1–4 23

b Inflorescence not a spike; perianth-segments 6 24

23a (22) Stamens 6 or more; carpels 3–6; perianth-segments 1–3, petal-like **224 Aponogetonaceae**

b Stamens 4; carpels 4; perianth-segments 4, not petal-like **226 Potamogetonaceae**

24a (22) Ovules many, borne parietally on the walls of the free carpels 25

b Ovules few, borne basally in each free carpel 26

25a (24) Leaves linear; latex absent; all perianth-segments petal-like **221 Butomaceae**

b Leaves with stalk and expanded blade; latex present; perianth differentiated into 3 green sepals and 3 coloured petals **221a Limnocharitaceae**

26a (24) Leaf-sheaths with ligules; flowers in racemes; perianth-segments all similar, greenish **223 Scheuchzeriaceae**

b Leaf-sheaths without ligules; flowers in whorls, racemes or panicles; perianth differentiated into sepals and petals **220 Alismataceae**

27a (20) Perianth differentiated into greenish calyx and petaloid corolla **28**

b Perianth-segments all similar, not differentiated into calyx and corolla or absent **30**

28a (27) Flowers solitary or in umbels; leaves broad, opposite or in a single whorl near the top of the flowering stem **229o Trilliaceae**

b Flowers in spikes, heads, cymes or panicles; leaves not as above **29**

29a (28) Stamens 6 or 3–5 with 1–3 staminodes; anthers basifixed; leaves usually borne on the stems, often with closed sheaths, never grey with scales; bracts neither overlapping nor conspicuously toothed **242 Commelinaceae**

b Stamens 6, staminodes 0; anthers dorsifixed; leaves mostly in basal rosettes, often rigid and spiny-margined, when on the stems then usually grey with scales; bracts usually overlapping and conspicuously toothed **241 Bromeliaceae**

30a (27) Plants obviously woody, at least at the bases of the stems, rarely composed of rosettes of fleshy leaves borne on woody rhizomes which are often concealed **31**

b Plants entirely herbaceous **38**

31a (30) Leaves reduced to scales, photosynthesis being carried out by flattened or divided stems (cladodes); fruit a berry **32**

b Leaves well-developed; fruit usually a capsule, more rarely a berry **33**

32a (31) Flowers borne directly on the cladodes **229q Ruscaceae**

b Flowers borne in inflorescences of various kinds directly on the stems, among the cladodes **229p Asparagaceae**

33a (31) Leaves succulent 34
b Leaves leathery or parchment-like or herbaceous but not succulent 35
34a (33) Flowers bell-shaped or almost spherical; perianth-segments united only at the extreme base **230 Agavaceae**
b Flowers tubular, at least the 3 outer perianth-segments united for more than half their length, the 3 inner segments united to the outer to varying degrees **229h Aloaceae**
35a (33) Leaves clearly stalked with 3–9 main veins arising at the top of the petiole clearly connected by cross-veinlets; climbers, rarely shrubs or creeping subshrubs 36
b Leaves not stalked, veins not as above, cross-veinlets obscure; shrubs or small trees 37
36a (35) Leaf-stalks bearing 2 tendrils; flowers inconspicuous; placentation axile **229s Smilacaceae**
b Leaf-stalks without tendrils; flowers very conspicuous, usually large; placentation often parietal **229r Philesiaceae**
37a (35) Flowers all bisexual, stalks not jointed; fruit a fleshy berry; ovules 1 in each cell of the ovary **230b Dracaenaceae**
b Flowers mostly unisexual, stalks jointed near their bases; ovules 2 in each cell of the ovary; fruit dry, indehiscent **230c Nolinaceae**
38a (30) Leaf-stalks bearing 2 tendrils; flowers foetid **229s Smilacaceae**
b Combination of characters not as above 39
39a (38) Leaves erect and blade-like, generally forming a tuft arising from a rhizome, at ground level, more rarely borne on a stem **230b Dracaenaceae**
b Combination of characters not as above 40
40a (39) Leaves reduced to small, scale-like spines; photosynthesis carried out by flattened or much-divided cladodes **229p Asparagaceae**
b Leaves well-developed, cladodes absent 41

41a (40) Plants with scapes, with small flowers without bracts in spikes or racemes; perianth calyx-like or absent; ovules 1 per cell, basal **225 Juncaginaceae**
b Combination of characters not as above 42
42a (41) Flowers solitary or in umbels; leaves broad, opposite or in a single whorl near the top of the flowering stem **229o Trilliaceae**
b Combination of characters not as above 43
43a (42) Flowers in umbels (rarely solitary, when subtended by 1 or more spathes) **229l Alliaceae**
b Flowers not in umbels, when solitary not subtended by spathes 44
44a (43) Plants emergent aquatics, either rooted in the substrate or free-floating 45
b Plants terrestrial 46
45a (44) Inflorescence a raceme subtended and enclosed in bud by a large spathe; leaves entire at the apex; anthers opening by slits **238 Pontederiaceae**
b Inflorescence not as above; leaves 2-toothed at the apex; anthers opening by pores **243 Mayacaceae**
46a (44) Placentation parietal or basal 47
b Placentation axile 48
47a (46) Leaves borne on the stems; flowers mostly unisexual, whitish **229n Asteliaceae**
b Leaves all basal, inflorescences borne on scapes; flowers all bisexual, usually yellow **244 Xyridaceae**
48a (46) Leaves equitant, distichous, forming fans; inflorescence a panicle **230d Phormiaceae**
b Combination of characters not as above 49
49a (48) Nectar secreted from glands on the outside of the ovary, generally in the grooves between the carpels 50
b Nectar secreted from glands on the perianth or rarely on the bases of the staminal filaments, rarely nectar absent 59

50a (49) Plants with bulbs **229k Hyacinthaceae**
b Plants without bulbs, usually with rhizomes **51**
51a (50) Leaves reduced to basal sheaths, without blades; perianth blue, flowers in heads **229d Aphyllanthaceae**
b Leaves with well-developed blades; perianth and flowers not as above **52**
52a (51) Leaves all basal; inflorescences borne on scapes (occasionally the scapes may bear small bracts) **53**
b At least some foliage leaves clearly borne on the flowering stems **56**
53a (52) Leaves fleshy and succulent; perianth-segments of the outer whorl united for part of their length, those of the inner whorl variably joined to the outer, all red, orange or yellow, tips often greenish **229h Aloaceae**
b Leaves not fleshy and succulent; perianth not as above **54**
54a (53) Leaves with a broadened blade and narrow, petiole-like base; flowers in a 1-sided raceme; perianth-segments fused below into a narrow tube, which broadens upwards **229f Hostaceae**
b Leaves without broadened blades and petiole-like bases; flowers in racemes, spikes or panicles, not as above; perianth-segments free or joined into a tube, which is not narrow below and broadened above **55**
55a (54) Perianth-segments united into a distinct tube below; stamens declinate **229e Hemerocallidaceae**
b Perianth-segments free or very slightly united at the base; stamens not declinate **229c Anthericaceae**
56a (52) Ovules 1 to many, axile from the upper inner angle of each cell of the ovary; sap often orange-red; perianth persistent in fruit, usually hairy outside **231 Haemodoraceae**
b Combination of characters not as above **57**
57a (56) Fruit a berry; perianth urceolate to bell-shaped or almost tubular, segments all united **229m Convallariaceae**

b Fruit a capsule; perianth various, not usually as above 58

58a (57) Carpels usually free above, styles usually 3; perianth-segments free **229a Melanthiaceae**

b Carpels united, style 1; perianth-segments free or united **229b Asphodelaceae**

59a (49) Style 1 sometimes divided above, or stigmas sessile; nectaries usually on the perianth-segments, rarely absent; plants usually with bulbs **229 Liliaceae**

b Styles usually 3, free to the base or almost so, more rarely 1; nectaries on the perianth-segments, on the bases of the staminal filaments or absent 60

60a (59) Style 1, ovary stalked; perianth-segments united; seeds hairy **229g Blandfordiaceae**

b Styles 3; ovary not stalked; perianth-segments free or if united then ovary borne below soil-level, in the corm; seeds not hairy 61

61a (60) Plants with corms; nectaries on the perianth or on the staminal filaments **229i Colchicaceae**

b Plants with rhizomes; nectaries absent **229a Melanthiaceae**

Group XV

Monocotyledons with the ovary inferior or half-inferior.

1a Ovary not completely inferior, generally about half above the point of attachment of the perianth and androecium 2

b Ovary totally inferior 4

2a (1) Anthers opening by pores **233 Tecophilaeaceae**

b Anthers opening by slits 3

3a (2) Styles 3; nectar secreted by glands on the perianth-segments **229a Melanthiaceae**

b Style 1; nectar secreted by glands on the the ovary, situated in the grooves between the carpels **229m Convallariaceae**

4a (1) Perianth actinomorphic or weakly zygomorphic; stamens 6, 4, 3 or rarely many, very rarely 5, 2 or 1 5 b Perianth strongly zygomorphic or asymmetric; stamens 5, 2 or 1 (very rarely 6), petaloid staminodes often present; if the perianth is weakly zygomorphic, then stamen 1, united with the style to form a column **19**

5a (4) Ovules 1–many, axile from the upper, inner angle of each cell of the ovary; perianth persistent in fruit, hairy outside; sap often orange **231 Haemodoraceae**

b Combination of characters not as above **6**

6a (5) Unisexual climbers with heart-shaped or very divided leaves; rootstock tuberous or woody **237 Dioscoreaceae**

b Combination of characters not as above **7**

7a (6) Rooted or floating aquatics; stamens 2–12; ovules distributed all over the carpel walls (placentation diffuse parietal) **222 Hydrocharitaceae**

b Terrestrial or marsh plants, or epiphytes; stamens 3 or 6, rarely many; placentation axile or parietal (when ovules restricted to a few well-defined placentas) **8**

8a (7) Stamens 3 **9**

b Stamens usually 6 or 3 + 3 staminodes, rarely many **10**

9a (8) Stamens on the same radii as the inner perianth-segments; flowers 3-winged from the ovary to the top of the perianth; leaves linear, flattened from top to bottom **239a Burmanniaceae**

b Stamens on the same radii as the outer perianth-segments; flowers not 3-winged; leaves usually equitant, flattened from side to side **239 Iridaceae**

10a (8) Placentation parietal; flowers in an umbel with the inner bracts long, thread-like and hanging **236 Taccaceae**

b Placentation usually axile; bracts and inflorescence not as above **11**

11a (10) Perianth consisting of an outer calyx-like whorl and an inner corolla-like whorl; bracts usually overlapping and conspicuously coloured **241 Bromeliaceae**

b Perianth not differentiated as above (though the segments of the 2 whorls may differ slightly); bracts not as above **12**

12a (11) Leaves long-persistent, evergreen, usually borne in a rosette, terminating in a spine or cylindrical body, more rarely not in a rosette but on creeping, slightly woody stems 13
b Leaves and stems not as above 15
13a (12) Plant a subshrub with creeping, woody stems **235 Velloziaceae**
b Plant forming a loose to dense rosette, stems scarcely apparent 14
14a (13) Leaves each terminating in a spine; flowers up to 6 cm long **230 Agavaceae**
b Leaves each terminating in a cylindrical body; flowers 10–15 cm long **230a Doryanthaceae**
15a (12) Flowers in a spike; leaves fleshy, often spotted with brown **230 Agavaceae**
b Flowers in heads, umbels or solitary 16
16a (15) Leaves hairy, pleated or with prominent veins **234 Hypoxidaceae**
b Leaves generally not as above 17
17a (16) Leaves reversed by an 180° twist in the narrow base; inflorescence a terminal umbel without spathes, with lateral flowering branches below it; seeds pale **229j Alstroemeriaceae**
b Leaves not reversed; flowers in a terminal umbel or solitary, subtended and enclosed in bud by 1–4 spathes; seeds black 18
18a (17) All leaves basal **232 Amaryllidaceae**
b At least some leaves borne on the flowering stems **232a Ixioliriaceae**
19a (4) Fertile stamens 6; perianth-segments all similar, united below into a curved and unevenly swollen tube; stem below ground, fleshy **230 Agavaceae**
b Stamens 5, 2 or 1, very rarely 6; staminodes, which may be petal-like, often present; perianth-segments usually differing among themselves; fleshy underground stems rare 20

20a (19) Fertile stamens 2 or 1, united with the style to form a column; pollen usually borne in coherent masses (pollinia); leaf-veins, when visible, all parallel to margins **258 Orchidaceae**

b Fertile stamens 5 or 1, rarely 6, not united with the style; pollen granular; leaf with a distinct midrib more or less parallel to the margins, the secondary veins parallel to each other, running at an angle from the midrib to the margins **21**

21a (20) Fertile stamens 5 or rarely 6 **22**

b Fertile stamen 1, petal-like staminodes 5 **25**

22a (21) Leaves and bracts spirally arranged; flowers unisexual; fruit a banana **254 Musaceae**

b Leaves and bracts in 2 ranks; flowers bisexual; fruit not a banana **23**

23a (22) Cymes arising from the bases of the leaf-sheaths; sepals united below into a long, stalk-like tube; median (upper) perianth-segments forming a large lip **254a Lowiaceae**

b Flowers in coiled cymes in the axils of spathes; outer perianth-segments free or at most joined to the inner; no perianth-segment forming a lip **24**

24a (23) Perianth-segments free; ovary with numerous ovules in each cell **254b Strelitziaceae**

b Perianth-segments partially united; ovary with 1 ovule per cell **254c Heliconiaceae**

25a (21) Fertile stamen with a thread-like filament and wider anther of 2 pollen-bearing lobes, not petal-like **26**

b Fertile stamen in part petal-like, with only 1 pollen-bearing anther-lobe **27**

26a (25) Leaves spirally arranged; lower apparent perianth-segment (the lip) formed from 5 united staminodes **255a Costaceae**

b Leaves distichous; lower apparent perianth-segment (lip) formed from 2 joined staminodes **255 Zingiberaceae**

27a (25) Leaf-stalk with a swollen band (pulvinus) at the junction with the blade; ovary smooth, with 1–3 ovules **257 Marantaceae** b Leaf-stalk without a pulvinus; ovary usually warty, with numerous ovules **256 Cannaceae**

Appendix

Key to dicotyledonous plants for which there are only male flowers (without obvious ovary rudiment or pistillode) available.

Plants with male flowers in catkins can be identified by using the Group X key (above) and are not included here. This key is really for guidance only, as the identification of purely male plants can often be very difficult. Some couplets in the key, notably 26, are very uncertain, and some families (e.g. Euphorbiaceae) are so variable that they have to be keyed several times, generally at the end of a string of leads, where they act as a catch-all that may trap genera and species from other families that have not been considered here.

1a Plants wholly parasitic 2
b Plants free-living or hemi-parasitic 5
2a (1) Plants without chlorophyll, leaves reduced to scales 3
b Plants with chlorophyll, with well-developed leaves 4
3a (2) Stamens numerous **51 Rafflesiaceae**
b Stamens 1 **157 Cynomoriaceae**
4a (2) Two united bracteoles forming a cup-like structure just below the perianth **15 Loranthaceae**
b Such bracteoles absent **15a Viscaceae**
5a (1) Plants aquatic, mostly submerged 6
b Plants terrestrial, sometimes marginal to water 9
6a (5) Leaves much divided into narrow, thread-like segments 7
b Leaves entire, not at all divided 8
7a (6) Stamens 10 or more, anthers with appendaged connectives **46 Ceratophyllaceae**
b Stamens 8 or fewer, anthers without appendages **154 Haloragaceae**

8a (6) Leaves whorled **156 Hippuridaceae**
b Leaves opposite **194 Callitrichaceae**
9a (5) Climbers; leaves with the blade running into a whip-like structure, which terminates in a pitcher **64 Nepenthaceae**
b Combination of characters not as above 10
10a (9) Plants herbaceous 11
b Plants woody 22
11a (10) Climbers, either twining, or with tendrils 12
b Upright or prostrate plants 14
12a (11) Plant climbing by means of tendrils; stamens usually 3, one anther 1-celled **144 Cucurbitaceae**
b Plant twining; stamens not as above 13
13a (12) Stamens borne on the perianth **21 Basellaceae**
b Stamens free from the perianth **10a Cannabaceae**
14a (11) Waterside plants with large, rhubarb-like leaves **154a Gunneraceae**
b Combination of characters not as above 15
15a (14) Leaves pinnate **142 Datiscaceae**
b Leaves simple, sometimes deeply lobed or palmate 16
16a (15) Leaves oblique at the base **143 Begoniaceae**
b Leaves symmetrical at the base 17
17a (16) Leaves opposite below, alternate above, all with sheathing bases **155 Theligonaceae**
b Leaves not as above 18
18a (17) Stipules absent 19
b Stipules present though sometimes falling early (scars) 20
19a (18) Stamens 8; leaves usually deeply divided **154 Haloragaceae**
b Stamens 1–5; leaves entire, toothed or shallowly divided **23 Chenopodiaceae**
20a (18) Leaves palmate, leaflets toothed **10a Cannabaceae**
b Leaves not as above 21
21a (20) Stamens inflexed in bud, springing erect when the flower opens **11 Urticaceae**

b Stamens not as above **94 Euphorbiaceae**
22a (10) Plants climbing **23**
b Plants creeping or upright **24**
23a (22) Stamens united into a coherent mass **32 Schisandraceae**
b Stamens variously arranged but not as above
44 Menispermaceae
24a (22) Stipules present, though sometimes falling early (scars) **25**
b Stipules completely absent **31**
25a (24) Plants with milky sap **26**
b Plants without milky sap **27**
26a (25) Generally trees; stamens usually borne on the perianth
10 Moraceae
b Generally shrubs; stamens not borne on the perianth
94 Euphorbiaceae
27a (25) Leaves palmately lobed, the petiole-bases completely covering the axillary bud; bark scaling in plates
72 Platanaceae
b Combination of characters not as above **28**
28a (27) Leaves opposite on long shoots, alternate on short shoots; flowers stalkless, each consisting of a group of naked stamens in a leaf-axil **40 Cercidiphyllaceae**
b Leaves not as above; flowers variously disposed, if axillary, then not as above **29**
29a (28) Leaf-stalk bearing 2 nectaries; leaves large, to 15 cm × 15 cm, cordate at the base; filaments hairy **130 Flacourtiaceae**
b Combination of characters not as above **30**
30a (28) Shrubs; leaves opposite, evergreen; flowers in spikes
49 Chloranthaceae
b Combination of characters not as above **94 Euphorbiaceae**
31a (24) Low, creeping shrublets with heather-like leaves
169 Empetraceae
b Upright shrubs or trees or climbers, leaves not heather-like **32**
32a (31) Anthers opening by pores; disc very well developed; leaves opposite; petals often fringed or lobed **123 Elaeocarpaceae**

b Combination of characters not as above 33
33a (32) Stamens borne on the perianth; anthers opening by valves 34
b Stamens not borne on the perianth; anthers opening by slits 35
34a (33) Leaves opposite **34 Monimiaceae**
b Leaves alternate **73 Hamamelidaceae**
35a (33) Calyx and corolla both present 36
b Perianth of a single whorl 37
36a (35) Sepals 2, petals 4 in 2 pairs; plants succulent, spiny **25 Didieriaceae**
b Sepals 4–5, petals 4–5; plants neither succulent nor spiny **119 Icacinaceae**
37a (35) Leaves evergreen 38
b Leaves deciduous 41
38a (37) Leaves alternate 39
b Leaves opposite 40
39a (38) Staminal filaments united **30 Myristicaceae**
b Staminal filaments free **95 Daphniphyllaceae**
40a (38) Stamens 4 **118 Buxaceae**
b Stamens 8–12 **118a Simmondsiaceae**
41a (37) Free-living trees with milky sap **9 Eucommiaceae**
b Hemi-parasites without milky sap **14 Santalaceae**

'Spot' characters

'Spot' characters are features that occur in relatively few families, thus rendering their identification simpler than tracing them through the actual key. Note that the appearance of a family name under a particular character does not mean that *all* members of that family show the feature in question.

Plants totally parasitic, without chlorophyll

51 Rafflesiaceae, 157 Cynomoriaceae, 168a Monotropaceae, 206 Orobanchaceae, 239a Burmanniaceae.

Plants clearly parasitic on the above-ground parts of other plants, with chlorophyll (excluding hemi-parasites)

14 Santalaceae, 15 Loranthaceae, 15a Viscaceae, 36 Lauraceae, 189 Convolvulaceae, 192 Lennoaceae.

Plants with bulbs

229 Liliaceae, 229k Hyacinthaceae, 229l Alliaceae, 232 Amaryllidaceae

Plants trapping insects by means of sticky hairs, pitchers or traps borne on stems or leaves, the remains of the insects usually digested

Pitchers: 63 Sarraceniaceae, 64 Nepenthaceae, 75 Cephalotaceae. *Active traps*: 65 Droseraceae, 207 Lentibulariaceae.

Sticky hairs: 22 Caryophyllaceae, 65 Droseraceae, 79 Byblidaceae, 80 Roridulaceae, 207 Lentibulariaceae.

Plants climbing by means of tendrils (coiling, often spring-like, much branched or bearing adhesive discs)

64 Nepenthaceae, 66a Fumariaceae, 84 Fabaceae, 107 Sapindaceae, 121 Vitaceae, 134 Passifloraceae, 144 Cucurbitaceae, 187a Cobaeaceae, 201 Bignoniaceae, 219 Compositae, 229 Liliaceae, 229s Smilacaceae.

Sap coloured or milky, not watery and translucent

Woody: 9 Eucommiaceae, 10 Moraceae, 94 Euphorbiaceae, 106 Aceraceae, 136 Bixaceae, 136a Cochlospermaceae, 158 Alangiaceae, 175 Sapotaceae. *Herbaceous*: 66 Papaveraceae, 94 Euphorbiaceae, 184 Apocynaceae, 185 Asclepiadaceae, 189 Convolvulaceae, 221a Limnocharitaceae, 248 Araceae.

Leaf-bases consistently oblique

8 Ulmaceae, 143 Begoniaceae.

Base of leaf-stalk completely enclosing the axillary bud

72 Platanaceae.

Translucent or dark, often aromatic glands, seen clearly in the leaves when held against the light

2 Myricaceae, 36 Lauraceae, 62 Guttiferae, 96 Rutaceae, 147 Myrtaceae.

At least some of the inflorescence bracts very conspicuous, often overlapping and coloured, or striking in some other way

Overlapping and coloured: 186 Rubiaceae, 202 Acanthaceae, 241 Bromeliaceae, 254 Musaceae, 254a Lowiaceae, 254b

Strelitziaceae, 254c Heliconiaceae. *Modified into nectar-secreting pouches*: 61 Marcgraviaceae. *Thread-like, long and dangling*: 236 Taccaceae.

Inflorescence a corymb, the inner flowers fertile and with actinomorphic corollas, the outer often sterile and with larger, zygomorphic corollas

68 Cruciferae, 76c Hydrangeaceae, 164 Umbelliferae, 211 Caprifoliaceae.

Epicalyx present

82 Rosaceae, 125 Malvaceae, 190 Hydrophyllaceae.

Corolla with a narrow, more or less parallel-sided tube and spreading lobes, with 2 stamens borne in the corolla-tube, their anthers back to back

179 Oleaceae.

Corolla with a distinctive outgrowth (corona)

134 Passifloraceae, 232 Amaryllidaceae.

Stamens on the same radii as petals, usually as many as petals, occasionally fewer, or, if petals are absent, on radii alternating with the perianth-segments

Woody, climbing: 121 Vitaceae, 122 Leeaceae. *Woody, not climbing*: 42 Berberidaceae, 109 Sabiaceae, 109a Meliosmaceae, 114 Corynocarpaceae, 171 Theophrastaceae, 172 Myrsinaceae. *Herbaceous*: 18 Nyctaginaceae, 42 Berberidaceae, 82 Rosaceae, 173 Primulaceae, 174 Plumbaginaceae.

Stamens 4, made up of 2 groups of $\frac{1}{2}+1+\frac{1}{2}$ stamens

49 Chloranthaceae, 66a Fumariaceae.

Stamens much longer than the petals, numerous, brightly coloured, the most conspicuous parts of the flower

60 Caryocaraceae, 84b Mimosaceae, 147 Myrtaceae.

Anthers opening by pores or by very short, pore-like slits

Woody: 56 Actinidiaceae, 57 Ochnaceae, 80 Roridulaceae, 84a Caesalpiniaceae, 102 Tremandraceae, 103 Polygalaceae, 123 Elaeocarpaceae, 136a Cochlospermaceae, 150 Melastomataceae, 166 Clethraceae, 168 Ericaceae, 197 Solanaceae. *Herbaceous*: 79 Byblidaceae, 84a Caesalpiniaceae, 103 Polygalaceae, 150 Melastomataceae, 167 Pyrolaceae, 182 Gentianaceae, 197 Solanaceae, 232 Amaryllidaceae, 233 Tecophilaeaceae, 243 Mayacaceae.

Anthers opening by valves

34 Monimiaceae, 36 Lauraceae, 42 Berberidaceae, 73 Hamamelidaceae.

Pollen collected together in structured units (pollinia), and dispersed as such

84a Caesalpiniaceae (few), 84b Mimosaceae (few), 185 Asclepiadaceae, 258 Orchidaceae

The mature ovary (rarely single carpels) clearly open at the top

69 Resedaceae, 142 Datiscaceae.

Ovary made up of 4 (rarely 5 or 2) apparently distinct units, appearing as dome-shaped humps, the single style gynobasic, arising from the hollow between them

57 Ochnaceae, 96 Rutaceae, 191 Boraginaceae, 193 Verbenaceae, 195 Labiatae.

Arrangement and description of families

In general, the variation given in the descriptions is somewhat wider than that presented in the key, and many characters used in the latter have had to be omitted. In using the descriptions, the following features must be assumed for most species of a family, unless otherwise stated: milky sap absent, habit not succulent, parts of the flower free from each other, stamens not antepetalous, and anthers opening by longitudinal slits. The ptyxis, as far as it is known, is given for each family: this refers to leaves if they are undivided, to leaflets if the leaves are divided.

The families are listed in the order of the Melchior system (see p. 3). No attempt has been made to group them into suprafamilial units ('orders'), as these have no significance for practical identification.

The following points concerned with presentation should be noted.

Morphology

The oblique stroke (/) is used instead of 'or'; the letter '*n*' is used instead of 'many' or 'numerous' (i.e. more than 10 or 12). Abbreviations: K, calyx-segments or sepals; C, corolla-segments or petals; P, perianth-segments when these are undifferentiated; A, stamens; G, carpels. These letters are also used in the collective sense; for example, 'A antepetalous' means stamens antepetalous. Brackets are used to indicate that the segments of any particular whorl are united to each other; for example, C(5) means a corolla of 5 lobes united below into a cup or tube.

In the dicotyledons it is not directly indicated whether the ovary is superior or inferior; however, it is always stated whether or not the perianth and stamens are hypogynous, perigynous or epigynous, which gives the same information (see pp. 40–45). Many of the families with the petals united into a tube at the base are described as 'K hypogynous, CA perigynous'; this means that the stamens are borne on the corolla and the ovary is superior. In the monocotyledons the position of the ovary is always indicated.

In plants with inferior ovaries, the calyx-segments (sepals) are shown as free (i.e. unbracketed) if the segments are completely free above the point of their attachment on the ovary, and as united if they are united above this point. Accurate information about this is difficult to obtain.

Information about the inflorescence-type is always given, but, this, again, is difficult to condense; the information given here should not be regarded as a complete description of the range of inflorescence types found in any particular family. Six main types of inflorescence are referred to: racemes, spikes, panicles, cymes, umbels and heads. The terms 'clusters' or 'fascicles' are used when the precise nature of the inflorescence is not understood. 'Racemose' has been used when spikes, racemes or basically indeterminate panicles occur in the same family; 'cymose' has been used in a similar way to cover various types of determinate inflorescence.

Ovule number refers to the number of ovules in each free carpel (if the ovary is made up of free carpels) or to the ovary as a whole (when it is made up of united carpels), unless qualified by 'per cell'.

Geography

This is indicated in general terms for each family. The number of genera native to Europe and North America is also given for each, as is an indication of how many genera are to be found in common cultivation.

Lists of genera

The genera in each family that occur in north temperate areas, or in cultivation in those areas, are listed. This listing is intended to provide simple guidance as to which genera are included. For more details (synonymy, authorities, etc.) see the book by Brummitt mentioned in the introduction (p. 3) and in the bibliography.

General

Some synonyms and short comments are included, where appropriate. The permitted alternative family names (for the eight families whose traditionally used names do not end in -'aceae') are given, separated from the more familiar name by an oblique stroke, e.g. **164. Umbelliferae/Apiaceae**.

Subclass Dicotyledones

Cotyledons usually 2, lateral; leaves usually net-veined, with or without stipules, alternate, opposite or whorled; flowers with parts in 2s, 4s or 5s, or parts numerous; primary root-system (taproot) usually persistent, branched.

1 (1) Casuarinaceae. Woody, branches jointed. Leaves whorled, scale-like. Inflorescence a catkin. Flowers unisexual, P0(?), A1, G(2), naked; ovules 2, parietal; styles 2, free. Samaras in woody cones. *Australia to Malaysia*.

Four genera, with about 80 species. A few of the 70 species of *Casuarina* are occasionally grown as ornamental shrubs.

Genera included: ***Casuarina***.

2 (2) Myricaceae. Woody. Leaves alternate, entire/divided, exstipulate, aromatic, gland-dotted; ptyxis conduplicate. Inflorescence a catkin, flowers unisexual. P0, A2–20, usually 4–8,

G(2), naked; ovule 1, basal; styles 2, free. Drupe. *Northern hemisphere.*

A family of 3 genera with about 30 species; 2 genera native in North America, 1 in Europe. A few species cultivated as ornamental, aromatic shrubs.

Genera included: ***Comptonia, Myrica.***

3 (3) Juglandaceae. Woody. Leaves usually opposite, pinnately compound, exstipulate; ptyxis (of leaflets) conduplicate. At least the male flowers in catkins. Flowers unisexual, P4, A3–*n*, G(2–3), inferior, ovule 1, basal; styles 2, free, sometimes divided, stigmas internal, lateral. Nut with complex lobed and folded cotyledons. *North temperate areas.*

Eight genera with about 50 species. Two genera are native to Europe (with 1 introduced) and 2 are native to North America. Four genera are cultivated as ornamental trees, and species of *Juglans* (walnut) and *Carya* (pecan) are cultivated for their edible nuts.

Genera included: ***Carya, Juglans, Platycarya, Pterocarya.***

4 (4) Leitneriaceae. Woody. Leaves alternate, entire, exstipulate. Inflorescence a catkin. Flowers unisexual, variously interpreted; male: P0, A3–12; female: P3–8, G1-celled, superior; ovule 1, lateral; style 1, stigma lateral. Drupe. *USA.*

The single genus, *Leitneria*, has only a single species; it occurs in the USA and is rarely cultivated in Europe.

Genera included: ***Leitneria.***

5 (5) Salicaceae. Woody, dioecious. Leaves usually alternate, simple, stipulate; ptyxis involute (*Populus*) or supervolute (*Salix*). Inflorescence a catkin, male catkins often erect. Flowers unisexual with disc or nectary-gland. P0, A2–*n*, G(2–4), superior; ovules *n*, parietal; stigmas 2–4 on a short style, or sessile. Capsule; seeds woolly. *Widespread.*

Two genera, *Salix* with more than 300 species and *Populus* with about 40 species. Both genera are native to North America and Europe, and species of both are cultivated as ornamental trees and shrubs.

Genera included: ***Populus, Salix***.

6 (6) Betulaceae. Trees or rarely small shrubs. Leaves alternate, simple, stipulate; ptyxis conduplicate-plicate. Flowers unisexual. Male flowers in pendent catkins, generally 3 to a bract, P4, A2/4; female flowers in pendent catkins or erect 'cones': perianth 0, G(2) ovary naked. Small nut. *Mostly north temperate areas*.

Two genera, *Betula* (60 species) and *Alnus* (15 species) both native in Europe and North America, and both widely cultivated as ornamental trees and shrubs.

Genera included: ***Alnus, Betula***.

6a (7) Corylaceae. Trees or small shrubs. Leaves alternate, simple, stipulate; ptyxis conduplicate-plicate. Flowers unisexual. Male flowers in usually pendulous catkins, P0, A4–15; female flowers in catkins or short spikes, P irregularly lobed, G(2), ovary inferior. Nut, sometimes enveloped in a cupule. *North temperate areas*.

There are 4 genera, of which 3 are native in Europe and 3 in North America. Species of all 4 are grown as ornamentals.

Genera included: ***Carpinus, Corylus, Ostrya, Ostryopsis***.

7 (8) Fagaceae. Woody; leaves usually alternate, simple, stipulate; ptyxis conduplicate, conduplicate-plicate or supervolute. Flowers unisexual. Male inflorescence a catkin; male flowers, P(4–7), A4–*n*; female flowers in clusters or catkins, P(4–7), G(3–6), inferior; ovules 2 per cell, axile; styles 3–6, free. Nut enveloped in a cupule. *Temperate and tropical areas*.

Eight genera and about 600 species. Three genera are native to Europe, 5 to North America. Species of many genera are important as timber trees, and those of 7 genera are cultivated as ornamentals. Includes Nothofagaceae.

Genera included: ***Castanea, Castanopsis, Chrysolepis, Fagus, Lithocarpus, Nothofagus, Quercus.***

8 (9) Ulmaceae. Woody. Leaves alternate, simple, stipulate, usually with oblique bases; ptyxis conduplicate. Flowers solitary/clustered, unisexual/bisexual, zygomorphic: PA hypogynous. P(4–9), A4–9, G(2); ovule 1, apical; styles 2, free, sometimes divided above. Samara/drupe. *Northern hemisphere.*

A family of 16 genera and about 140 species. Three genera native to Europe, 5 to North America. Species of 3 genera are grown as ornamental trees and shrubs. Includes Celtidaceae.

Genera included: ***Celtis, Planera, Trema, Ulmus, Zelkova.***

9 (10) Eucommiaceae. Trees with milky sap. Leaves alternate, simple, exstipulate; ptyxis supervolute. Flowers solitary, unisexual, actinomorphic. P0, A4–10, G(2) naked; ovule 1, apical; styles 2, free. Samara. *China.*

A single genus and species (*Eucommia ulmoides*). It is grown as an ornamental and has some interest in that its milky sap contains rubber.

Genera included: ***Eucommia.***

10 (11) Moraceae. Woody plants, usually with milky sap. Leaves alternate/opposite, simple/divided, stipulate; ptyxis conduplicate or supervolute. Inflorescence various, flowers sometimes sunk in expanded receptacles, which may take the form of hollow cups. Flowers unisexual. PA hypogynous. Male P2–6, usually 4, A1–5; female P2–6 or entire and enveloping ovary, G(2), often 1 carpel aborting; ovule 1, apical; styles1–2, free. Syncarps. *Tropical and north temperate areas.*

There are about 50 genera and up to 1500 species.

Genera included: ***Artocarpus, Brosimum, Broussonetia, Chlorophora, Cudrania, Dorstenia, Fatoua, Ficus, Maclura, Morus, Pseudolmedia, Streblus, Trophis.***

10a (12) Cannabaceae. Herbs or non-woody climbers without milky sap. Leaves alternate/opposite, simple/divided, stipulate. Inflorescence a raceme or spike. Flowers unisexual. PA hypogynous. Male: P(5), A1–5; female: P enveloping ovary, G(2); ovule 1, apical; styles 1–2 free. Achene. *Temperate Eurasia.*

Two genera. *Humulus* (hops) is native to Europe; *Cannabis*, now widely cultivated all over the world, often illegally for its narcotic resin or legally for fibre, probably originated in central Asia.

Genera included: ***Cannabis, Humulus.***

11 (13) Urticaceae. Herbs/rarely shrubs, often with rough or stinging hairs. Leaves alternate/opposite, simple, usually stipulate, ptyxis conduplicate or involute. Inflorescence various. Flowers unisexual, actinomorphic. PA usually hypogynous/ovary naked. P0–5/(2–5), A3–5, usually 4 inflexed in bud and touch-sensitive; G1; ovule 1, basal; style 1/rarely 0, stigma often brush-like. Achene/drupe. *Widespread.*

There are about 50 genera and 2000 species. Five genera occur in Europe and 13 in North America. Species of several genera are cultivated as fibre-plants, and of about 8 as ornamentals.

Genera included: ***Boehmeria, Debregeasia, Elatostema, Forsskaolea, Hesperocnide, Laportea, Myriocarpa, Neraudia, Parietaria, Pilea, Pipturus, Pouzolzia, Roussellia, Soleirolia, Urera, Urtica.***

12 (14) Proteaceae. Trees/shrubs. Leaves alternate, exstipulate, evergreen, often very hard; ptyxis flat or conduplicate.

Inflorescence various. Flowers usually bisexual, actinomorphic/zygomorphic. PA perigynous. P(4), A4 rarely 1–3 infertile, borne on petal-like, spoon-shaped P-segments, G1; ovules 1–*n*, marginal; style 1, thickened or with pollen-collecting apparatus at the apex. Follicle/nut/drupe. *Southern hemisphere.*

There are about 75 genera and up to 1300 species. Many are highly ornamental, though difficult to grow; species of about 16 genera are in cultivation.

Genera included: ***Aulax, Banksia, Bellendena, Embothrium, Gevuina, Grevillea, Hakea, Knightia, Leucadendron, Leucospermum, Lomatia, Macadamia, Protea, Roupala, Stenocarpus, Telopea.***

13 (15) Olacaceae. Woody, sometimes half-parasitic on the roots of other trees. Leaves alternate, simple, exstipulate. Flowers solitary/in clusters/racemes/panicles, bisexual, actinomorphic. KCA hypogynous. K4–6, C3–5/(6), A3+5 staminodes/5/1–10, G(3) usually 1-celled; ovules 2 or 3, axile; style 1; stigma 2–3-lobed. Drupe. *Mostly tropics.*

A family of about 27 genera and 320 species. Two genera occur in North America, and a few are cultivated as ornamentals or curiosities.

Genera inlcuded: ***Dulacia, Heisteria, Olax, Ongokea, Ptychopetalum, Schoepfia, Ximenia.***

14 (16) Santalaceae. Herbs/shrubs/trees, parasitic on the roots of other plants. Leaves alternate/opposite, simple, exstipulate. Inflorescence spike/raceme/cluster/flowers solitary. Flowers unisexual/bisexual, actinomorphic. PA epigynous/rarely half-epigynous. P3–5/(3–5), A3–5, G(3–5); ovules 1–5, basal; style 1, stigma 2–5-lobed. Nut/drupe. *Widespread.*

Thirty-five genera with about 400 species; 3 genera occur in Europe, 8 in North America. Species of perhaps 6 genera are occasionally grown as ornamentals.

Genera included: ***Buckleya, Comandra, Exocarpos, Nestronia, Osyris, Pyrularia, Santalum, Thesium.***

15 (17) Loranthaceae. Woody/herbaceous branch parasites. Leaves opposite/whorled, simple, exstipulate, usually evergreen. Flowers unisexual, actinomorphic, solitary/in pairs, each flower with a cup-like structure formed from 2 bracteoles just below the perianth. PA epigynous. P4–6/(4–6), A2–6, G(3–6); ovules not differentiated in flower; styles absent, stigma sessile, unlobed. Berry/drupe, 2–3-seeded. *Mostly tropics.*

About 50 genera with up to 1000 species. There are 2 genera native to North America, 1 to Europe, and species of all 3 of these are cultivated.

Genera included: ***Dendropemon, Loranthus.***

15a (18) Viscaceae. Similar to the *Loranthaceae*, but leaves and stems sometimes absent, flowers without a cup-like structure just beneath the perianth. *Widespread.*

There are 12 genera and about 400 species. Five genera are native to North America, 2 to Europe. Species of *Viscum* (mistletoe) are cultivated, especially for Christmas decoration.

Genera included: ***Arceuthobium, Dendrophthora, Korthalsella, Phoradendron, Viscum.***

16 (19) Polygonaceae. Herbs/shrubs/climbers/rarely trees. Leaves alternate or all basal, simple/lobed, stipules usually present, often united into a sheath (ochrea); ptyxis revolute. Flowers in racemes/ cymes, unisexual/bisexual, actinomorphic. P usually hypogynous. P3–6/rarely (3–6), A6–9, G(2–4), usually (3); ovule 1, basal; styles 2–4, free/occasionally slightly united at base. Nut. *Mainly north temperate areas.*

There are about 30 genera with some 800 species. Twenty-two genera are native to North America, 10 to Europe, and species of about 10 are cultivated as ornamentals.

Genera included: ***Antigonon, Atraphaxis, Brunnichia, Calligonum, Centrostegia, Chorizanthe, Coccoloba, Emex, Eriogonum, Fagopyrum, Fallopia, Gilmania, Goodmania, Hollisteria, Homalocladium, Koenigia, Lastarriaea, Mucronea, Muehlenbeckia, Nemacaulis, Oxyria, Oxytheca, Persicaria, Polygonella, Polygonum, Reynoutria, Rheum, Rumex, Stenogonum.***

17 (20) Phytolaccaceae. Trees/shrubs (some climbing)/herbs. Leaves alternate, entire, exstipulate; ptyxis conduplicate. Flowers in racemes, usually bisexual, actinomorphic. PA hypogynous. P4–5/(4–5), A3–*n*, G1–(*n*)/rarely *n*; ovules 1/1 per cell, basal/axile. Fruit often fleshy. *Mainly American tropics and southern hemisphere.*

There are about 20 genera and 100 species. One genus is introduced into Europe and 5 are native to North America; species of 4 genera are cultivated as ornamentals. *Petiveria* is sometimes separated off as the Petiveriaceae.

Genera included: ***Ercilla, Microtea, Petiveria, Phytolacca, Rivina, Trichostigma.***

17a (21) Agdestidaceae. Woody climber with large, swollen rootstock. Leaves alternate, cordate, exstipulate. Flowers in panicles, bisexual, actinomorphic. PA half-epigynous. P4–5, A15–30, G(3–4), semi-inferior; style 1, divided above. Achene retained in the persistent perianth.

A single genus, *Agdestis*, native to the southern part of the USA and Central America; introduced further south.

Genera included: ***Agdestis.***

17b (22) Achatocarpaceae. Dioecious shrubs or small trees, with thorny shoots. Leaves alternate, entire, exstipulate. Racemes/panicles. Flowers unisexual, PA hypogynous. P4–5,

A10–20, G(2), 1-celled with 1 basal ovule; styles 2. Berry. *Southern USA to south America.*

Two genera, one of them, *Phaulothamnus*, native to the southern USA.

Genera included: ***Phaulothamnus***.

17c (23) Molluginaceae. Herbs, often annual and ephemeral. Leaves alternate or in false whorls, simple, usually exstipulate. Flowers bisexual in axillary cymes. PA hypogynous. P4–5/(4–5), A3–5/*n*, G(2–5), ovules *n*, axile; styles 2–5, free or very slightly joined at the base. Capsule. *Widespread but very scattered.*

There are about 12 genera; species of *Mollugo* and *Glinus* are found in Europe.

Genera included: ***Glinus, Mollugo***.

17d (24) Gisekiaceae. Herbs. Leaves opposite or in false whorls, simple, exstipulate. Flowers bisexual in axillary cymes. PA hypogynous. P5, A5–20, G5/rarely more, each carpel with 1 ovule and an individual style. Cluster of dimorphic achenes, some smooth, some tuberculate-spiny. *Tropical and subtropical Africa and Asia, introduced elsewhere.*

There is a single genus, *Gisekia*, with about 5 species.

Genera inlcuded: ***Gisekia***.

18 (25) Nyctaginaceae. Herbs/shrubs/woody climbers. Leaves alternate/opposite/whorled, usually opposite, entire, exstipulate; ptyxis conduplicate. Inflorescence cymose. Flowers usually bisexual, actinomorphic. PA hypogynous. P(5), A1-*n*/(1-*n*) usually 5/(5), G1-celled; ovule 1, basal; style 1, stigmas slightly divided at apex. Achene, often in persistent P. *Mostly tropics.*

There are 30 genera and about 300 species: 3 genera (2 introduced) occur in Europe, whereas 16 genera are native in North America. Species and hybrids of a few genera, notably *Mirabilis*

(Marvel of Peru) and *Bougainvillea*, are widely grown as ornamentals.

Genera inlcuded: ***Abronia, Acleisanthes, Allionia, Allionella, Ammocodon, Anulocaulis, Boerhavia, Bougainvillea, Commicarpus, Guapira, Mirabilis, Neea, Nyctaginia, Okenia, Oxybaphus, Pisonia, Selinocarpus, Tripterocalyx.***

19 (26) Aizoaceae. Herbs/shrubs, usually leaf-succulents. Leaves usually opposite, sometimes all basal, simple, exstipulate. Flowers in cymes, bisexual, actinomorphic. KCA epigynous. K4–15/(4–15), Cn, An, G(3–20); ovules usually n, parietal/rarely axile; styles free or stigmas sessile, radiating. Fruit usually a complex capsule opening actively when wetted, closing when dry/rarely fleshy and indehiscent. *Desert areas, mainly South Africa.*

There are about 120 genera and some 2500 species, mostly from South Africa and adjacent areas; most of the genera were originally part of the large genus *Mesembryanthemum*. Species of many genera are grown in cultivation as ornamental succulents. Includes Mesembryanthemaceae.

Genera included: ***Acrodon, Aizoon, Aloinopsis, Amphibolia, Antegibbaeum, Aptenia, Argyroderma, Astridia, Bergeranthus, Bijlia, Braunsia, Carpanthea, Carpobrotus, Caryotophora, Cephalophyllum, Chasmatophyllum, Cheiridopsis, Conicosia, Conophytum, Cylindrophyllum, Dactylopsis, Delosperma, Didymaotus, Dinteranthus, Disphyma, Dorotheanthus, Drosanthemum, Eberlanzia, Ebracteola, Faucaria, Fenestraria, Frithia, Galenia, Gibbaeum, Glottiphyllum, Hereroa, Herrea, Herreanthus, Jacobsenia, Juttadinteria, Kensitia, Lampranthus, Lapidaria, Leipoldtia, Lithops, Machairophyllum, Malephora, Mesembryanthemum, Meyerophytum, Mitrophyllum, Monilaria, Neohenricia, Odontophorus, Oophytum, Ophthalmophyllum, Oscularia, Pleisopilos,***

Rhinephyllum, Rhombophyllum, Ruschia, Schwantesia, Sesuvium, Stoeberia, Stomatium, Tetragonia, Titanopsis, Trianthema, Trichodiadema, Vanheerdia, Vanzijlia.

20 (27) Portulacaceae. Herbs/shrubs, often fleshy. Leaves alternate/opposite, simple, entire stipulate/exstipulate; ptyxis very variable. Inflorescence racemes/cymes/rarely flowers solitary. Flowers bisexual, actinomorphic. KCA hypogynous/partially epigynous. K2/rarely 3 or more, C3–18/(3–18), when united only so at the extreme base, usually 4–6, A3–*n*, on the same radii as petals when few, G(2–8), 1-celled; ovules 1–*n*, basal/free-central; styles 2–8, very slightly united below/rarely single. Capsule/rarely indehiscent. *Mostly New World.*

Nineteen genera with about 500 species. Two genera are native to Europe and 8 to North America. Species of 9 genera are cultivated as ornamentals and *Portulaca oleracea* is grown as a salad.

Genera included: ***Anacampseros, Calandrinia, Calyptridium, Claytonia, Lewisia, Montia, Portulaca, Portulacaria, Spraguea, Talinopsis, Talinum.***

21 (28) Basellaceae. Climbers. Leaves alternate, simple, fleshy, exstipulate. Inflorescence racemose. Flowers unisexual/bisexual, actinomorphic. PA perigynous. P5/(5), A5, G(3), 1-celled; ovule 1, basal; style 1, stigmas usually 3. Drupe in persistent, fleshy P. *Mostly tropical America.*

There are 4 genera and 17 species. Two genera occur in North America and some are cultivated for ornament.

Genera included: ***Anredera, Basella, Boussingaultia, Ullucus.***

22 (29) Caryophyllaceae. Herbs/rarely shrublets. Leaves usually opposite, simple, entire, exstipulate/rarely stipulate;

ptyxis flat, conduplicate or rarely supervolute. Inflorescence cymose/flowers solitary. Flowers usually bisexual, actinomorphic. KCA hypogynous. K4–5/(4–5), C4–5, A8–10/rarely fewer, G(2–5); ovules usually *n*, free-central; styles 2–5, free. Capsule/rarely fleshy. *Mostly north temperate areas.*

A family of about 90 genera and more than 2000 species. Over 30 genera are native in Europe, and over 30 in North America. The family is one of the most important in providing ornamental herbaceous plants, especially species of *Dianthus* (pinks, carnations), *Saponaria* (soapwort), *Silene, Gypsophila*, etc.; species of about 20 genera are found in gardens.

Genera included: ***Agrostemma, Alsinidendron, Arenaria, Bolanthus, Bufonia, Cerastium, Colobanthus, Cucubalus, Dianthus, Drymaria, Drypis, Gypsophila, Holosteum, Honckenya, Loeflingia, Lychnis, Minuartia, Moerhringia, Moenchia, Myosoton, Ortegia, Petrocoptis, Petrorhagia, Polycarpon, Pseudostellaria, Sagina, Saponaria, Schiedea, Silene, Spergula, Spergularia, Stellaria, Stipulicida, Telephium, Vaccaria, Velezia, Wilhelmsia.***

22a (30) Illecebraceae. Herbs. Leaves usually opposite, usually stipulate, stipules often hyaline and shining. Flowers axillary, small, bisexual. PA/rarely KCA perigynous. P5/rarely K5, C5, A5, often with 5 staminodes, G(2–3), superior; ovule 1, basal; style 1, slightly divided above. Nut.

About 12 genera, scattered in north temperate areas. As accepted here, the family is defined by the perigynous perianth and stamens and the 1-seeded, indehiscent fruit. Other interpretations are current.

Genera included: ***Achyronychia, Cardionema, Corrigiola, Geocarpon, Herniaria, Illecebrum, Paronychia, Pteranthus, Scleranthus, Scopulophila.***

23 (31) Chenopodiaceae. Herbs/shrubs, often succulent. Leaves alternate/opposite, usually simple, exstipulate, reduced to scales when stems fleshy and segmented. Inflorescence usually cymose. Flowers unisexual/bisexual, actinomorphic. PA usually hypogynous. P(3–5)/rarely 0, green/membranous, A usually 5, G(2–3), 1-celled, rarely half-inferior; ovule 1, basal; styles 2–5, usually free. Achene/nut. *Widespread.*

About 100 genera and 1400 species. Thirty-four genera occur in Europe and 25 in North America. Very few are cultivated, though species from about 10 genera can be found in European gardens, mostly as foliage plants. One species of *Beta* is grown for its sugar-yielding roots.

Genera included: ***Agriophyllum, Allenrolfea, Anabasis, Aphanisma, Arthrocnemum, Atriplex, Axyris, Bassia, Beta, Bienertia, Camphorosma, Ceratocarpus, Chenopodium, Corispermum, Cycloloma, Girgensohnia, Grayia, Halimione, Halimocnemis, Halocnemum, Halogeton, Halopeplis, Halostachys, Haloxylon, Horaninovia, Kalidium, Kochia, Krascheninnikovia, Microcnemum, Nanophyton, Nitrophila, Noaea, Ofaiston, Petrosimonia, Polycnemum, Salicornia, Salsola, Sarcobatus, Sarcocornia, Spinacia, Suaeda, Suckleya, Zuckia.***

24 (32) Amaranthaceae. Herbs/woody. Leaves alternate/opposite/whorled, usually entire, exstipulate; ptyxis flat or conduplicate. Inflorescence often racemose and very condensed. Flowers usually bisexual, actinomorphic. PA hypogynous. P3–5/(3–5), usually hyaline and/or papery, A usually (5), staminodes frequent, G(2–3); ovules 1–*n*, basal; style 1, stigma slightly bilobed/styles 2–3. Capsule/achene/berry. *Mostly tropics.*

Sixty genera and 900 species are known, 4 genera occurring in Europe, 18 in North America. A few (especially species of *Amaranthus*) are in cultivation.

Genera included: ***Achyranthes, Aerva, Alternanthera, Amaranthus, Bosea, Celosia, Chamissoa, Froelichia, Gomphrena, Guilleminia, Hermbstaedtia, Iresine, Lithophila, Nototrichium, Ptilotus, Pupalia, Tidestromia.***

25 (33) Didieriaceae. Woody, succulent, often spiny. Leaves alternate, exstipulate. Cymes/panicles/umbels. Flowers unisexual, actinomorphic. KCA hypogynous. K2, persistent, C4 in 2 pairs, A usually 8/10, G(3–4), 1-celled; ovule 1, basal; style 1, stigma 3–4-lobed. Fruit dry, dehiscent. *Madagascar*.

Four genera and 11 species; a few are grown as curiosities.

Genera included: ***Alluaudia, Decarya, Didieria.***

26 (34) Cactaceae. Mostly spiny stem-succulents. Leaves usually absent, when present ptyxis supervolute (*Pereskia*). Flowers usually solitary and bisexual, actinomorphic/slightly zygomorphic. KCA usually epigynous. K*n*, C*n*/(*n*), A*n* G(3–*n*); ovules n, parietal; style 1, stigma much divided at apex. Berry. *Mostly New World*.

There are over 100 genera and at least 1500 species, all from the New World except for one, which occurs in Africa, Madagascar and Sri Lanka. Nineteen genera occur in North America. Many genera and species are cultivated.

Genera included: ***Acanthocereus, Ariocarpus, Armatocereus, Arrojadoa, Astrophytum, Aztekium, Browningia, Calymmanthium, Carnegeia, Cephalocereus, Cereus, Cleistocactus, Coleocephalocereus, Copiapoa, Corryocactus, Coryphantha, Denmoza, Discocactus, Disocactus, Echinocactus, Echinocereus, Echinopsis, Epiphyllum, Epithelantha, Escobaria, Espostoa, Eulychnia, Ferocactus, Frailea, Gymnocalycium, Haageocereus, Harrisia, Hatiora, Hylocereus, Lepismium, Leuchtenbergia, Lophophora, Maihuenia, Mammillaria, Melocactus, Mila, Myrtillocactus, Neolloydia, Neoporteria, Neoraimondia, Newowerdermannia, Obregonia, Opuntia, Oreocereus, Pachycereus, Parodia, Pediocactus, Pelecyphora, Peniocereus, Pereskia, Pereskiopsis, Pilosocereus, Pterocactus, Rebutia, Rhipsalis, Schlumbergera, Sclerocactus, Selenicereus, Stenocactus, Stenocereus, Stetsonia, Strombocactus, Thelocactus, Uebelmannia, Weberocereus.***

27 (35) Magnoliaceae. Trees/shrubs. Bark aromatic. Leaves simple, deciduous/evergreen, alternate with large, deciduous stipules which enclose buds; ptyxis conduplicate. Flowers bisexual, actinomorphic, solitary. PA hypogynous. P in several series, in 3s or 4s (usually 6 or 9), A*n*, spirally arranged, G*n*, spirally arranged; ovules 2–*n*, marginal; styles 1 per carpel. Fruit a group of follicles; seeds large, with arils. *Mostly north temperate and subtropical areas.*

A family of 12 genera and about 200 species. Two genera are native in North America, and species of 3 genera (especially *Magnolia*) are widely cultivated.

Genera included: ***Liriodendron, Magnolia, Manglietia, Michelia***.

28 (36) Winteraceae. Woody. Leaves alternate, simple, evergreen, exstipulate; ptyxis supervolute (*Drimys*). Flowers unisexual/bisexual, actinomorphic, in cymes/fascicles. KCA hypogynous. K2–6/(2–6), valvate, C2–*n* in 2–several series, A*n*, G1–*n* in 1 whorl; ovules 1–*n*, marginal; styles 1 per carpel, free. Follicle/berry. *Tropics (except Africa), south temperate areas.*

Possibly 6 genera and about 80 species. A few species of *Drimys* and *Pseudowintera* are cultivated as ornamental shrubs.

Genera included: ***Drimys, Pseudowintera***.

29 (37) Annonaceae. Woody. Leaves simple, deciduous/evergreen, exstipulate; ptyxis conduplicate. Inflorescence various. Flowers usually bisexual, actinomorphic. KCA hypogynous. K usually 3, C3–6/P6–9, A*n*, each crowned with an enlarged connective, G*n*, usually stalked in fruit, rarely united into a mass; ovules 1–*n*, basal/marginal; styles 1 per carpel, free. Berry or aggregate of berries; seeds with arils, endosperm convoluted. *Tropics, temperate areas in the New World.*

A large family of 120 genera and about 2100 species, mainly tropical, but with 6 genera native to North America. Species of about 6 genera are cultivated.

Genera included: ***Annona, Artabotrys, Asimina, Cananga, Guatteria, Monodora, Oxandra, Rollinia, Xylopia.***

30 (38) Myristicaceae. Dioecious trees. Leaves alternate, entire, evergreen, exstipulate. Inflorescence racemose. Flowers unisexual, actinomorphic. PA hypogynous. P(2–5), usually (3), valvate in bud; male: A(2–10), filaments united; female: G1, ovule 1, basal; style 1. Fruit fleshy, 2-valved; seed with coloured aril, endosperm convoluted. *Tropics.*

There are 15 genera and about 250 species. Species of *Myristica* and *Pycnanthus* are occasionally found in cultivation (in glasshouses) and *Myristica fragrans* is the source of nutmeg and mace (seed and aril, respectively).

Genera included: ***Myristica, Pycnanthus.***

31 (39) Canellaceae. Woody, bark very aromatic. Leaves alternate, simple, exstipulate. Inflorescence cymes/racemes. Flowers bisexual, actinomorphic. KCA hypogynous. K3, C4–5, A(10+), filaments united, G(2–6), 1-celled; ovules 2–*n*, parietal; style 1, short, stigma 2–6-lobed. Berry. *Tropical and subtropical Africa and America.*

A family of 6 genera and about 20 species; 2 genera are native to North America and species of *Canella* are sometimes cultivated.

Genera included: ***Canella, Pleodendron.***

32 (40) Schisandraceae. Woody climbers, monoecious or dioecious. Leaves alternate, simple, exstipulate; ptyxis involute (*Kadsura*) or supervolute (*Schisandra*). Flowers unisexual, actinomorphic, axillary. PA hypogynous. P5–20/K3, C2–20, A(*n*) usually united into a fleshy mass, G*n*; ovules 2–3 per carpel, marginal; styles free, 1 per carpel. Fruit berry-like, crowded or distant on an elongate axis. *North America, East Asia.*

A family of 2 genera and about 45 species. One genus (*Schisandra*) is native in North America, and species of both genera are cultivated.

Genera included: ***Kadsura, Schisandra.***

33 **(41) Illiciaceae.** Woody, aromatic. Leaves alternate/whorled, evergreen, simple, exstipulate; ptyxis supervolute. Flowers bisexual, actinomorphic, solitary. PA hypogynous. P7–*n*, in several whorls, imbricate, the inner often larger than the outer, A4–*n*, G5–*n* in a single whorl; ovule 1 per carpel, almost basal; styles 1 per carpel, free. Fruit a group of follicles.

A family of a single genus with about 40 species; 2 of the species are native to North America and several are cultivated as aromatic, ornamental shrubs.

Genera inlcuded: ***Illicium.***

34 **(42) Monimiaceae.** Trees/shrubs, aromatic. Leaves opposite, simple, exstipulate, usually evergreen; ptyxis conduplicate. Flowers solitary/cymose, usually unisexual, actinomorphic. PA perigynous. P4–*n*, occasionally in 2 whorls but all similar, A usually *n*, each with a pair of glands or cup-like appendages at base, G1–*n*; ovule 1 per carpel, basal; styles free, 1 per carpel. Achenes in enlarged perigynous cup. *Mainly tropics.*

A family of 34 genera and about 450 species. A few species of *Atherosperma, Laurelia* and *Peumus* are grown as ornamental trees and shrubs in glasshouses. Includes *Atherospermataceae.*

Genera inlcuded: ***Atherosperma, Laurelia, Peumus.***

35 **(43) Calycanthaceae.** Shrubs, aromatic. Leaves opposite, entire, deciduous/rarely evergreen, exstipulate; ptyxis flat-conduplicate. Flowers solitary, bisexual, actinomorphic. PA perigynous. P*n*, A5–30, G*n*; ovules 1–2 per carpel; styles 1 per carpel, free. Achenes in persistent perigynous zone. *China, south-eastern USA.*

Two genera with up to 10 species. One genus (*Calycanthus*) is native to North America, and species of it and of *Chimonanthus* (China) are cultivated as early-flowering shrubs.

Genera included: ***Calycanthus, Chimonanthus.***

36 (44) Lauraceae. Woody, aromatic. Leaves usually alternate, entire, exstipulate, usually evergreen, glandular-punctate; ptyxis conduplicate or supervolute. Inflorescence cymose/racemose. Flowers small, unisexual/bisexual, actinomorphic. PA hypogynous/perigynous. P usually (4/6), A8–12/variable, anthers opening by valves; G1; ovule 1, apical; style 1. Drupe-like berry. *Mostly tropics.*

There are nearly 50 genera and about 2000 species, mostly tropical but with 2 genera occurring in Europe (1 introduced) and 13 native to North America. Species of a few genera are grown as ornamental shrubs. *Laurus nobilis* is the bay, used as a flavouring, and *Persea indica* produces the avocado pear. *Cassytha* is parasitic.

Genera included: ***Aniba, Beilschmiedia, Cassytha, Cinnamomum, Laurus, Licaria, Lindera, Litsea, Nectandra, Neolitsea, Ocotea, Persea, Phoebe, Sassafras, Umbellularia.***

37 (45) Tetracentraceae. Woody. Leaves alternate, simple, exstipulate. Inflorescence catkin-like. Flowers bisexual, actinomorphic. PA hypogynous. P4, A4, G(4); ovules *n*, apical; styles 4, free. Capsule. *China and adjacent Burma, Himalaya.*

A family of a single genus and species (*Tetracentron sinense*), commonly grown in European gardens.

Genera included: ***Tetracentron.***

38 (46) Trochodendraceae. Woody. Leaves whorled, simple, exstipulate, evergreen; ptyxis supervolute. Flowers actinomorphic, bisexual, racemose/in clusters. PA hypogynous. P minute/0, A*n*, G(6–10) in a single whorl; ovules *n*, marginal; styles 6–10, free. Fruit a group of coalesced follicles. *Japan, Korea.*

Like the previous family, of a single genus and species (*Trochodendron aralioides*), which is widely cultivated.

Genera included: ***Trochodendron***.

39 (47) Eupteleaceae. Woody. Leaves alternate but in false whorls, deciduous, exstipulate; ptyxis conduplicate. Flowers bisexual;, actinomorphic, in the leaf-axils. P0, A*n*, hypogynous, G6–*n*, stalked; ovules 1–3 per carpel, marginal; styles 1 per carpel, free. Fruit a group of stalked samaras. *Himalaya, Japan, China.*

A family of a single genus and 2 very similar species, occasionally grown as small ornamental trees.

Genera included: ***Euptelea***.

40 (48) Cercidiphyllaceae. Woody, dioecious. Leaves deciduous, simple, opposite on long shoots, alternate on short shoots, stipulate; ptyxis involute. Flowers unisexual. Male: almost stalkless, P4, A15–20; female: stalked, P4, hypogynous, G4–6; ovules *n*, marginal; styles 4–6, free. Follicles. *China, Japan.*

One genus (*Cercidiphyllum*) with a single species (occasionally regarded as 2 species), cultivated as a handsome small tree.

Genera inlcuded: ***Cercidiphyllum***.

41 (49) Ranunculaceae. Usually herbs/rarely woody/rarely climbing. Leaves usually alternate, simple/compound, usually exstipulate; ptyxis variable, mainly conduplicate or supervolute. Inflorescence various. Flowers bisexual, actinomorphhic/zygomorphic/KCA/PA hypogynous. P4–*n*/K3–5, C2–*n*/ rarely (4), bearing nectaries, A5–10/*n*, G1–*n*/rarely carpels united; ovules 1–*n*, marginal/basal/apical, rarely axile; styles 1 per carpel, long or short, free. Achenes/follicles/rarely berry-like. *Mainly temperate areas.*

About 50 genera and over 2000 species. Twenty-three genera are native in Europe and 24 in North America. The family provides many ornamental herbaceous plants, and some popular woody climbers (*Clematis*). *Hydrastis* is sometimes separated off as the Hydrastidaceae.

Genera included: ***Aconitum, Actaea, Adonis, Anemone, Anemonella, Anemonopsis, Aquilegia, Callianthemum, Caltha, Ceratocephala, Cimicifuga, Clematis, Consolida, Coptis, Delphinium, Eranthis, Helleborus, Hepatica, Hydrastis, Isopyrum, Myosurus, Nigella, Paraquilegia, Pulsatilla, Ranunculus, Semiaquilegia, Thalictrum, Trautvetteria, Trollius, Xanthorhiza.***

41a (50) Glaucidiaceae. Rhizomatous perennial herbs. Leaved palmately lobed, mostly basal but 2 on each flowering stem, alternate, exstipulate. Flowers terminal, solitary, actinomorphic. PA hypogynous. P4, A*n*, G(2), carpels united only at the extreme base, ovules *n*, axile, styles 2, free. Follicles flattened laterally, somewhat 4-winged. *Japan.*

A single genus (*Glaucidium*) with a single species; often cultivated.

Genera included: ***Glaucidium.***

42 (51) Berberidaceae. Herbs/shrubs. Leaves alternate/rarely opposite, simple/divided, usually exstipulate, evergreen/deciduous; ptyxis variable. Inflorescence cymose/racemose/flowers solitary. Flowers bisexual, actinomorphic. KCA hypogynous, K and C sometimes not well differentiated. K3-*n*/rarely 0, C4–6/rarely 9/rarely 0, bearing nectaries, A4–18, often on the same radii as the petals, anthers opening by valves, G apparently 1; ovules few, basal/marginal; style 1, short. Capsule/berry. *Mainly north temperate areas.*

Sometimes divided into several further families, including Berberidaceae in the narrow sense, Nandinaceae,

Diphyllaeiaceae, Leonticaceae and Podophyllaceae. In the broad sense used here, there are 15 genera and 570 species; 6 genera (1 introduced) are found in Europe and 10 in North America. Many genera, notably *Berberis, Mahonia* and *Epimedium*, are very ornamental and are widely cultivated.

Genera included: ***Achlys, Berberis, Bongardia, Caulophyllum, Diphylleia, Epimedium, Gymnospermium, Jeffersonia, Leontice, Mahonia, Nandina, Podophyllum, Ranzania, Vancouveria.***

43 (52) Lardizabalaceae. Woody, monoecious or dioecious, usually climbers. Leaves alternate, compound, exstipulate; ptyxis conduplicate. Inflorescence racemose. Flowers usually unisexual, actinomorphic. PA hypogynous. P6/9, sometimes in 2 whorls, A 6/(6), G3–10; ovules *n* per carpel, marginal; styles 1 per carpel, stigmas sessile. Berries. *Scattered.*

A family of 8 genera and 21 species. One genus is native in North America. Species of about 5 genera are cultivated.

Genera included: ***Akebia, Decaisnea, Holboellia, Lardizabala, Sinofranchetia, Stauntonia.***

44 (53) Menispermaceae. Usually woody climbers. Leaves alternate, simple, deciduous/evergreen, exstipulate; ptyxis flat or conduplicate. Flowers unisexual, actinomorphic, in racemes/panicles. PA/KCA hypogynous. K4–*n*, C6–9/P3/6/9, A3/6/9/*n*, G3–6; ovules 1 per carpel, marginal; styles 3–6, free. Drupe/achene; seeds weakly to conspicuously horseshoe-shaped. *Mostly tropics.*

There are 67 genera and 470 species. Five genera are native in North America, and species of some 7 genera are grown as interesting climbers.

Genera included: ***Anamirta, Calycocarpum, Chondrodendron, Cissampelos, Cocculus, Dioscoreophyllum, Hyperbaena, Jateorhiza, Menispermum, Sinomenium.***

45 (54) Nymphaeaceae. Rhizomatous aquatic herbs; leaves and flowers usually floating, rarely raised above the water-surface. Leaves alternate, cordate/peltate, prickly or not beneath; ptyxis involute. Flowers bisexual, actinomorphic. KCA serially attached to the ovary (which therefore can be partly inferior). K4–6, C*n*, A*n*, G(3–35), superior to inferior; ovules *n* per cell, parietal; stigmas sessile on top of the ovary. Fruit various. *Widespread.*

There are about 9 genera with 100 species. Two genera are native both to Europe and to North America. Plants with leaves prickly on the under-surface (*Euryale, Victoria*) are sometimes separated off as the Euryalaceae.

Genera included: ***Euryale, Nuphar, Nymphaea, Victoria.***

45a (55) Cabombaceae. Leaves both submerged and floating, the submerged leaves opposite or whorled, very finely divided, the floating leaves entire (occasionally absent), peltate, without sinuses. Flowers solitary, floating or raised above the surface, KCA hypogynous. K3, C3, A3–6, G2–4, superior; ovules usually 3 per carpel, pendulous. *North and South America, Africa, Northeast Asia.*

There are 2 genera; *Cabomba* is native to Atlantic North America.

Genera included: ***Brasenia, Cabomba.***

45b (56) Nelumbonaceae. Leaves alternate, floating when young, held well above the water when mature, peltate, widely funnel-shaped, without sinuses. Flowers held well above the water. KCA hypogynous. K4–5, C10–25, A*n*, each appendaged, G12–20, free and sunk individually in a top-shaped receptacle; ovule 1 per carpel, apical. *Warm temperate and tropical America and Asia.*

Nelumbo is the sole genus; *N. lutea* is native to the eastern and southern parts of the USA.

Genera included: ***Nelumbo.***

46 (57) Ceratophyllaceae. Submerged aquatic herbs. Leaves whorled, much divided, exstipulate. Flowers solitary, unisexual, actinomorphic. PA hypogynous. P(8–13), A10–20, connectives prolonged at the top, G1; ovule 1, marginal; style single. Nut. *Widespread.*

One genus, native to both Europe and North America.

Genera included: ***Ceratophyllum***.

47 (58) Saururaceae. Herbs. Leaves alternate, simple, stipulate; ptyxis involute. Inflorescence spike/raceme/head. Flowers bisexual, actinomorphic. P0, A6–8, G3–5/(3–5), superior/inferior, 1-celled or 3–5-celled, ovules 1–10 per cell, parietal/axile; styles 3–4, free. Follicle/fleshy capsule. *Scattered.*

There are 4 genera and about 6 species. Two genera (*Saururus* and *Anemopsis*) are native to North America.

Genera included: ***Anemopsis, Houttuynia, Saururus***.

48 (59) Piperaceae. Herbs/shrubs. Leaves usually alternate, entire, stipulate/exstipulate; ptyxis variable. Inflorescence a fleshy spike. Flowers minute, usually bisexual, often sunk in the spike. P0, A2–10, G(2–4), superior; ovule 1, basal; stigma sessile, often brush-like. Small drupe. *Tropics.*

A family of 8 genera and over 3000 described species. Three genera occur in North America and species of 2 (*Peperomia, Piper*) are widely cultivated as ornamentals. The fruits of *Piper nigrum* are the source of pepper.

Genera included: ***Peperomia, Piper, Pothomorphe***.

49 (60) Chloranthaceae. Herbs/shrubs. Leaves opposite, simple, stipulate; ptyxis conduplicate. Inflorescence spike/panicle/head. Flowers unisexual, actinomorphic; male A1–3, sometimes made up of a central stamen attached to 2 half-stamens; female P3, epigynous, G 1-celled; ovule 1, pendulous; stigma sessile. Drupe. *Tropics, south temperate areas.*

Five genera with about 40 species. One genus is native in the southern parts of North America, and one or two are occasionally grown as ornamentals.

Genera included: ***Chloranthus, Hedyosmum, Sarcandra***.

50 (61) Aristolochiaceae. Herbs/climbers. Leaves alternate, simple, often cordate, exstipulate; ptyxis conduplicate. Inflorescence various. Flowers bisexual, actinomorphic/zygomorphic. PA epigynous. P(3) often bizarre in shape and foetid, A6 often attached to style, G(4–6); ovules *n*, axile; stigmas several, sessile. Capsule. *Mostly tropics.*

There are 7 genera and about 600 species. Two genera are native to Europe and 3 to North America. Species of *Asarum* and *Aristolochia* are grown for their bizarre flowers.

Genera included: ***Aristolochia, Asarum, Hexastylis***.

51 (62) Rafflesiaceae. Root- or branch-parasites lacking chlorophyll. Leaves scale-like/0. Flowers unisexual, actinomorphic. PA epigynous. P4–10(4–10), A*n*, G(4/6/8); ovules *n*, parietal. Berry. *Mostly Old World tropics.*

There are 9 genera (1 native to Europe) and about 55 species. The flowers of *Rafflesia* (Malaysia, Indonesia) are the largest known. *Cytinus* is sometimes separated off as the Cytinaceae.

Genera included: ***Cytinus***.

52 (63) Dilleniaceae. Usually woody, sometimes climbing/herbaceous. Leaves alternate, simple, exstipulate; ptyxis conduplicate. Inflorescences various. Flowers bisexual, actinomorphic. KCA hypogynous. K5/4–18 enlarging in fruit, C5, A*n* sometimes in 2–5 bundles, G1–20, sometimes somewhat coherent; ovules 1–*n*, marginal; styles free. *Mostly tropics, Australasia.*

A family of about 12 genera and 150 species. One genus is native to North America; species of 2 others are occasionally cultivated for ornament.

Genera included: ***Dillenia, Doliocarpus, Hibbertia***.

53 (64) Paeoniaceae. Herbs/soft-wooded shrubs. Leaves alternate, compound, exstipulate; ptyxis variable. Flowers usually solitary, bisexual, actinomorphic. KCA hypogynous. K5, persistent and differing in size, C5–9, A*n*, G2–8; ovules *n*, marginal; styles free; nectar-secreting disc present below the ovary. Large follicles with coloured seeds. *North temperate areas.*

A single genus with about 25 species, native to both Europe and North America, and widely cultivated.

Genera included: ***Paeonia***.

54 (65) Crossosomataceae. Shrubs. Leaves alternate, simple, exstipulate. Flowers solitary, terminal, bisexual, actinomorphic. KCA perigynous. K5, C5, A15–*n*, G3–5; ovules *n*, marginal; styles free. Follicles; seeds with much-divided arils. *Western North America.*

A family of 4 genera, with 9 species.

Genera included: ***Apacheria, Crossosoma, Glossopetalon.***

55 (66) Eucryphiaceae. Trees/shrubs, evergreen in the wild, tending to be deciduous in cultivation. Leaves opposite, simple/pinnate, stipulate (stipules falling early); ptyxis revolute. Flowers solitary, bisexual, actinomorphic. KCA hypogynous. K(4), falling as a unit as flower opens, C4, A*n*, G(5–12 or rarely more); ovules *n*, axile; styles free, short. Capsule. *Australia (Tasmania), Chile.*

A further family of a single genus with about 10 species; several species and hybrids are extensively cultivated as extremely ornamental, late-flowering small trees or shrubs.

Genera included: ***Eucryphia***.

56 (67) Actinidiaceae. Woody, some climbing. Leaves alternate, simple, evergreen/deciduous, exstipulate; ptyxis supervolute or conduplicate. Inflorescences various. Flowers unisexual/bisexual, actinomorphic. KCA hypogynous. K3–8, usually

5, C3–8, usually 5, A*n*, anthers opening by pores, G(3–5); ovules *n*, axile; styles usually free, occasionally joined at base. Berry/capsule. *Subtropics and tropics.*

There are 3 genera and about 100 species; species of all 3 are cultivated for ornament. *Actinidia chinensis* is widely grown in warm areas (especially New Zealand) for its edible berries (kiwi fruit).

Genera included: ***Actinidia, Clematoclethra, Saurauia.***

57 (68) Ochnaceae. Trees/shrubs. Leaves alternate, simple, stipulate, mostly evergreen. Flowers bisexual, actinomorphic. KCA hypogynous. K5, C5–10, A*n*, anthers opening by slits or apical pores, G3–*n*, united only by a common, single style, which is slightly lobed at the top; ovules 1–*n* per cell, axile. Schizocarp, often fleshy. *Mostly tropics.*

A family of 37 genera and about 450 species. Two genera are native to North America and a few species of *Ochna* are cultivated as ornamental greenhouse shrubs. *Lophira* is sometimes separated off as the Lophiraceae.

Genera included: ***Lophira, Ochna, Ouratea, Sauvagesia.***

58 (69) Dipterocarpaceae. Very large trees. Leaves alternate, entire, evergreen, stipulate; ptyxis conduplicate. Inflorescence usually a panicle. Flowers bisexual, actinomorphic. KCA hypogynous. K5, 2 usually enlarging in fruit, C5, A*n*, G usually (3); ovules usually 2 per cell, axile/rarely parietal. Nut. *Tropical Asia, Africa, America.*

A family of 16 genera and some 530 species, all of them very large tropical trees; juvenile specimens of a few species are grown in glasshouses in Europe, where they rarely flower.

Genera included: ***Shorea.***

59 (70) Theaceae. Trees/shrubs/rarely climbing. Leaves alternate, simple, exstipulate, usually evergreen; ptyxis conduplicate or supervolute. Inflorescence raceme/panicle/flowers solitary.

Flowers bisexual, actinomorphic. KCA hypogynous/K hypogynous CA perigynous. K5–6, C5–14, A*n*, filaments sometimes partly united below, united to bases of petals, G(3–6); ovules *n*, axile; styles 3–6, free/united and single, lobed at the top. Capsule/berry. *Tropics and warm temperate areas.*

A family of about 30 genera and 520 species. Seven genera are native in North America, and species of several others are widely cultivated as ornamental shrubs. The most commonly grown is *Camellia*; the tea plant, widely grown in subtropical areas, belongs to this genus (*C. sinensis*). Includes *Ternstroemiaceae.*

Genera included: ***Camellia, Cleyera, Eurya, Franklinia, Gordonia, Laplacea, Schima, Stuartia, Ternstroemia, Visnea.***

60 (71) Caryocaraceae. Woody. Leaves alternate/opposite, evergreen, of 3 leaflets, stipulate/exstipulate. Inflorescence a raceme. Flowers bisexual, actinomorphic. KCA hypogynous. K(5–6), C5–6/rarely (5–6), A*n*, with long, projecting, coloured filaments, G(4–20); ovules 1 per cell, axile; styles 4–20, free. Drupe/schizocarp. *Tropical America.*

Two genera and 24 species; very uncommon in cultivation.

Genera included: ***Caryocar.***

61 (72) Marcgraviaceae. Woody climbers, often epiphytic. Leaves alternate, simple, exstipulate, evergreen; ptyxis supervolute. Inflorescences racemes/umbels; flowers bisexual, zygomorphic, some sterile with their bracts variously modified into pitcher-like, pouched or spurred nectaries. KCA hypogynous. K4–7/(4–7), C4–5/(4–5), falling as a unit when united, A3–*n*, G(2–*n*); ovules n, parietal; style single, short, or stigmas sessile. Capsule/indehiscent. *Tropical America.*

There are 5 genera and about 100 species. One genus is native in North America, and species of *Marcgravia* are occasionally cultivated as evergreen woody climbers with bizarre inflorescences.

Genera included: ***Marcgravia.***

62 (73) Guttiferae/Clusiaceae. Herbs/woody, mostly with yellow, clear or blackish resinous latex contained in discrete glands. Leaves usually opposite/rarely whorled/alternate, usually exstipulate, gland-dotted; ptyxis flat or conduplicate. Inflorescence cymose. Flowers unisexual/bisexual, actinomorphic. KCA hypogynous. K2–6/(2–6), C2–10, A*n*, often united in 3–5 bundles, G(2–10)/1-celled; ovules 1–*n* per cell, axile; styles several, free/united below/rarely stigmas sessile. Capsule/berry. *Widespread.*

There are 48 genera and over 1000 species. The family Hypericaceae, consisting of herbs or shrubs mainly from temperate areas, is often distinguished; it is native to both North America and Europe.

Genera included: ***Calophyllum, Clusia, Garcinia, Hypericum, Mammea, Triadenium.***

63 (74) Sarraceniaceae. Herbs. Leaves basal tubular pitchers, exstipulate. Flowers solitary/racemose, bisexual, actinomorphic. KCA/PA hypogynous. K4–6, C5/0, A8–*n*, G(3–5); ovules n, axile; style 1, peltately expanded. Capsule. *North and South America.*

There are 3 genera, of which 2 are native to North America; species of all 3 are cultivated.

Genera included: ***Darlingtonia, Heliamphora, Sarracenia.***

64 (75) Nepenthaceae. Woody climbers. Leaves alternate, exstipulate, the tip of the blade prolonged into a stalked insectivorous pitcher. Inflorescence a raceme/panicle. Flowers unisexual, actinomorphic. PA hypogynous. P3–4, A*n*, G(3–4); ovules *n*, axile; stigmas sessile. *Tropical southeast Asia.*

A single genus with about 70 species; several of these, and hybrids between them, are widely cultivated.

Genera included: ***Nepenthes.***

65 (76) Droseraceae. Herbs/small shrubs. Leaves in basal rosettes, usually simple, insectivorous with sticky hairs or traps, exstipulate, ptyxis circinate. Flowers bisexual, actinomorphic, KCA hypogynous. K4–5, C4–5, A4–5/10–20, pollen in tetrads, G(3–5); ovules *n*, parietal; styles 3–5 free/single, lobed at apex. Capsule. *Temperate areas.*

Four genera with well over 100 species. Three of the genera are native to Europe and 2 to North America. Species of all four are cultivated. *Drosophyllum* is sometimes separated off as the Drosophyllaceae.

Genera included: ***Aldrovanda, Dionaea, Drosera, Drosophyllum.***

66 (77) Papaveraceae. Herbs/rarely shrubs, with milky or coloured sap. Leaves alternate/basal/rarely opposite, simple/divided, exstipulate; ptyxis variable. Inflorescence cymose/flowers solitary. Flowers bisexual, actinomorphic (though with 2 obvious planes of symmetry). KCA hypogynous/rarely perigynous. K2–3/rarely 4–*n*/rarely (2) and falling as a whole, always falling as flower opens, C0–*n*, usually 4/6, A*n*, G(2–*n*), rarely almost free; ovules *n*, parietal; style 1 or styles very short, stigmas several, sessile or almost so; nectar absent. Capsule. *Mostly north temperate areas.*

About 23 genera with over 300 species, many genera native to both Europe and North America, and many are cultivated (especially *Papaver* and *Meconopsis*). The family is economically important because of the opium poppy (*Papaver somniferum*), from which various drugs are extracted (morphine, codeine).

Genera included: ***Arctomecon, Argemone, Bocconia, Canbya, Chelidonium, Dendromecon, Dicranostigma, Eomecon, Eschscholtzia, Glaucium, Hesperomecon, Hunnemannia, Hylomecon, Macleaya, Meconella, Meconopsis, Papaver, Platystemon, Roemeria, Romneya, Sanguinaria, Stylomecon, Stylophorum.***

66a (78) Fumariaceae. Herbs, occasionally climbing; sap clear. Leaves alternate, divided, exstipulate. Inflorescence racemose. Flowers bisexual, often with 2 planes of symmetry/zygomorphic. KCA hypogynous. K2, persistent, C 4 in 2 pairs, the bases of 1 or both the 2 outer often spurred, the 2 inner simple/tripartite and spreading/erect with the apices joined into a pollen sac, which encloses the anthers, A4, sometimes made up of 2 groups of 2 half-stamens united to a whole stamen; G(2), style often elaborate; ovules 1–*n*, parietal; seeds often arillate; nectar secreted at the bases of the stamens, collecting in the corolla spurs (if present). *North temperate areas.*

Eighteen genera; 8 occur in Europe and about 5 in North America. Many species of *Corydalis* and *Dicentra* are cultivated. This family is included in the Papaveraceae by many authors (including Brummitt). Two genera, *Pteridophyllum* (C4, all more or less the same, unspurred, A4, leaves all basal, fern-like) and *Hypecoum* (C4, the 2 outer diamond-shaped, the 2 inner tripartite, A4) are sometimes separated off as 2 further families, respectively Pteridophyllaceae and Hypecoaceae.

Genera included: ***Adlumia, Ceratocapnos, Corydalis, Dicentra, Fumaria, Hypecoum, Platycapnos, Pteridophyllum, Rupicapnos, Sarcocapnos.***

67 (79) Capparaceae. Woody/herbaceous. Leaves alternate, simple/compound, stipulate/exstipulate; ptyxis conduplicate. Flowers solitary/in racemes, somewhat zygomorphic, bisexual. KCA hypogynous. K4–8, usually 4, C4–8, usually 4, A4/6–*n*, G(2) often stalked; ovules few–*n*, parietal; style short, stigma bilobed/sessile. Capsule/berry/nut. *Tropics, warm temperate areas.*

A family of 45 genera and about 700 species. Two genera are native to Europe and 9 to North America. Species of a few genera are cultivated as ornamentals. The buds of *Capparis spinosa* provide the capers of commerce. *Cleome* and its allies are sometimes separated off as the Cleomaceae, and *Koeberlinia* as the Koeberliniaceae.

Genera included: ***Atisquamea, Capparis, Cleome, Cleomella, Crateva, Koeberlinia, Morisonia, Oxystylis, Polanisia, Wislizenia.***

68 (80) Cruciferae/Brassicaceae. Mostly herbs. Leaves usually alternate, simple/divided, exstipulate, rarely with basal lobes appearing like stipules; ptyxis variable. Inflorescence a spike/raceme, often corymbose, generally without bracts. KCA hypogynous/rarely perigynous. K4, C4/rarely 0, A usually 6, 2 shorter and 4 longer/rarely fewer or more, G(2); ovules 2–*n*, parietal, but with a false septum dividing the ovary cell after fertilisation; stigmas sessile or style single, short. Usually capsule dehiscing by valves separating from the false septum/rarely indehiscent. *Cosmopolitan in cooler areas.*

A large and rather uniform family of 390 genera and 3000 species. One hundred and eight genera are native to Europe and 94 to North America. Many are grown as ornamentals, others as vegetable crops (species of *Brassica*), salads (*Eruca, Lepidium, Nasturtium*) or spices (*Sinapis*).

Genera included: ***Aethionema, Alliaria, Alyssoides, Alyssum, Anastatica, Andrzeiowskia, Anelsonia, Aphragmus, Arabidopsis, Arabis, Armoracia, Athysanus, Aubrieta, Aurinia, Barbarea, Berteroa, Biscutella, Bivonaea, Boleum, Bornmuellera, Brassica, Braya, Bunias, Cakile, Calepina, Camelina, Capsella, Cardamine, Cardaminopsis, Carrichtera, Caulanthus, Caulostramina, Chlorocrambe, Chorispora, Christolea, Chrysocamela, Clausia, Clypeola, Cochlearia, Coincya, Coluteocarpus, Conringia, Coronopus, Crambe, Degenia, Descurainia, Didesmus, Dimorphocarpa, Diplotaxis, Diptychocarpus, Dithyrea, Draba, Dryopetalon, Enarthrocarpus, Erophila, Eruca, Erucaria, Erucastrum, Erysimum, Euclidium, Eutrema, Euzomodendron, Fibigia, Glaucocarpum, Goldbachia, Guiraoa, Halimolobos,***

Heliophila, Hesperis, Heterodraba, Hirschfeldia, Hornungia, Hugueninia, Hymenolobus, Iberis, Idahoa, Iodanthus, Ionopsidium, Isatis, Kernera, Leavenworthia, Lepidium, Lepidotrichum, Leptaleum, Lesquerella, Lobularia, Lunaria, Lycocarpus, Lyrocarpa, Malcolmia, Mancoa, Maresia, Matthiola, Megacarpaea, Moricandia, Morisia, Murbeckiella, Myagrum, Neotorularia, Nerisyrenia, Neslia, Notoceras, Notothlaspi, Pachyphragma, Parrya, Peltaria, Pennellia, Petrocallis, Phoenicaulis, Physaria, Polyctenium, Pritzelago, Raffenaldia, Raphanus, Rapistrum, Rhizobotrya, Ricotia, Rorippa, Schivereckia, Schizopetalon, Schoenocrambe, Selenia, Sibara, Sinapis, Sisymbrella, Sisymbrium, Smelowskia, Sobolewskia, Stanleya, Sterigmostemum, Streptanthella, Streptanthus, Subularia, Succowia, Synthlipsis, Syrenia, Tauscheria, Teesdalia, Teesdaliopsis, Tetracme, Thellungiella, Thelypodiopsis, Thelypodium, Thlaspi, Thysanocarpus, Tropidocarpum, Vella, Warea, Wasabia.

69 (81) Resedaceae. Herbs/shrubs. Leaves alternate, simple/divided, stipules minute. Inflorescence a spike/raceme. Flowers unisexual/bisexual, zygomorphic. KCA usually hypogynous. K4–8, C2–8, usually fringed, A3–*n*, G(2–6)/rarely 2–6, open at the top; ovules *n*, parietal/marginal; styles absent, stigmas surrounding the open apex of the ovary. Capsule/follicle/berry, usually open and gaping at the apex.

Six genera with 75 species. Two genera are native to Europe, 2 to North America. A few species of *Reseda* are grown as ornamentals or for their scented inflorescences.

Genera included: ***Oligomeris, Reseda, Sesamoides.***

70 (82) Moringaceae. Trees. Leaves alternate, 2–3-pinnate, exstipulate though with glands at the base of the stalk. Flowers bisexual, zygomorphic, in panicles. KCA perigynous. K5, C4–5, A5 + 3–5 staminodes, anthers 1-celled, G(3); ovules *n*, parietal; style 1, long, slender. Capsule, 3-sided. *Mostly Old World tropics.*

A family of a single genus of about 12 species occurring in mainly desert areas. A few of the species are grown as ornamentals.

Genera included: ***Moringa***.

71 (83) Bataceae. Shrubs. Leaves opposite, simple, exstipulate. Inflorescence a catkin. Flowers unisexual; male: P2, A4–5 + 4–5 staminodes; female: P0, G(4), naked; ovules 1 per cell, ascending; styles free. Fruit a syncarp of berries. *Coasts of America.*

A family of a single genus with 2 species occurring on coastal regions of both North and South America.

Genera included: ***Batis***.

72 (84) Platanaceae. Trees with exfoliating bark, often with stellate hairs. Leaves alternate, palmately lobed, stipulate, leaf-bases covering the axillary buds; ptyxis conduplicate-plicate. Inflorescence a raceme of hanging, spherical heads. Flowers unisexual, actinomorphic. PA perigynous/hypogynous. P3–5/(3–5) sometimes considered as bracts, A3–4, G5–9; ovules 1/rarely 2 marginal; styles free. Fruit prickly balls of achenes with persistent styles. *North temperate areas.*

A family of a single genus with about 8 species, occurring in both Europe and North America. Several species and hybrids are cultivated, often as street trees.

Genera included: ***Platanus***.

73 (85) Hamamelidaceae. Woody, often with stellate hairs. Leaves usually alternate, simple/lobed, stipulate; ptyxis flat/conduplicate/rarely supervolute. Flowers in spikes, clusters or pairs, unisexual/bisexual, actinomorphic/zygomorphic. KCA perigynous/epigynous. K4–5/(4–5), C4–5/rarely fewer or more/P0–5, A4–5/rarely more, anthers opening by valves, G(2); ovules

1–*n* per cell, axile; styles 2, free. Woody capsule. *Tropics and subtropics, mainly E Asia, few temperate.*

A family of 28 genera and about 90 species. Three genera are native to North America. Species of several genera are cultivated as (often early-) flowering shrubs, especially in the genera *Corylopsis* and *Hamamelis*. *Rhodoleia* is sometimes separated off as the Rhodoleiaceae.

Genera included: ***Corylopsis, Disanthus, Distylium, Exbucklandia, Fortunearia, Fothergilla, Hamamelis, Liquidambar, Loropetalum, Parrotia, Parrotiopsis, Rhodoleia, Sinowilsonia, Sycopsis.***

74 (86) Crassulaceae. Herbs/shrubs, with succulent leaves. Leaves alternate/opposite, simple, exstipulate. Flowers in cymes/spikes/panicles/racemes/rarely solitary, bisexual, actinomorphic. KCA hypogynous. K4–7/12–16/(4–5), C4–7/12–16/(4–5), A4–5/10–12/rarely more, G4–5/10–12; ovules *n*, marginal; styles free. Follicles. *Widespread (except Australia).*

There are 30 genera and about 1400 species, all leaf-succulents; almost all of the genera are in cultivation. Thirteen genera are native to Europe, 12 to North America.

Genera included: ***Adromischus, Aeonium, Aichryson, Bryophyllum, Chiastophyllum, Cotyledon, Crassula, Cremnophila, Diamorpha, Dudleya, Echeveria, Graptopetalum, × Graptoveria, Greenovia, Jovibarba, Kalanchoe, Lenophyllum, Monanthes, Mucizonia, Orostachys, Pistorinia, Rhodiola, × Sedeveria, Sedum, Sempervivum, Sinocrassula, Thompsonella, Tylecodon, Umbilicus, Villadia.***

75 (87) Cephalotaceae. Herbs. Leaves alternate, in a rosette, modified into stalked, insectivorous pitchers, exstipulate. Flowers in racemes, bisexual, actinomorphic. PA perigynous. P6, A12, connectives swollen, glandular, G6; ovules 1 per carpel, basal/marginal; styles free. Follicles. *Australia.*

A single genus with a single species (*Cephalotus folicularis*), which is occasionally cultivated in Europe.

Genera included: ***Cephalotus.***

76 (88) Saxifragaceae. Herbs. Leaves alternate, often all basal, usually simple, sometimes evergreen, exstipulate/rarely with small stipules; ptyxis variable. Flowers solitary/in racemes, usually bisexual, actinomorphic/rarely zygomorphic. KCA perigynous/epigynous. K4–5, C4–5/rarely 0, A4–5/8–10, G(2)/rarely more, occasionally almost free; ovules *n*, axile; styles 2, free, usually divergent. Capsule. *Mainly north temperate areas.*

About 30 genera and 475 species, with genera native to both Europe and North America. Many of the genera are widely cultivated, especially species and hybrids of the largest genus, *Saxifraga*. The genus *Francoa*, with evergreen, usually divided leaves and flowers with parts in 4s, is sometimes separated off as a family (Francoaceae).

Genera included: ***Astilbe, Astilboides, Bensoniella, Bergenia, Bolandra, Boykinia, Chrysosplenium, Conimitella, Darmera, Elmera, Francoa, Heuchera, ×Heucherella, Jepsonia, Leptarrhena, Lithophragma, Mitella, Mukdenia, Peltoboykinia, Rodgersia, Saxifraga, Suksdorfia, Sullivantia, Tanakea, Tellima, Tiarella, Tolmiea.***

76a (89) Penthoraceae. Perennial herbs. Leaves alternate, exstipulate. Flowers bisexual, actinomorphic, in terminal cymes. PA perigynous. P5, somewhat unequal, A10, G(5), united only at the base; styles 5, ovules numerous. Fruit a group of follicles. *China, North America.*

A family of a single genus with 2 species, one of which is native to North America.

Genera included: ***Penthorum.***

76b (90) Parnassiaceae. Herbs. Leaves mostly basal, simple, exstipulate. Flowers solitary, actinomorphic, bisexual. KCA hypogynous. K5/(5), C5, A5 alternating with 5 multifid staminodes, G(3–4); ovules *n*, parietal; style absent, stigmas more or less sessile. Fruit a capsule. *Arctic and north temperate areas.*

A family with 2 genera, one native to both Europe and North America, the other to North America. Both are difficult in cultivation and rarely seen. *Lepuropetalon* is sometimes separated off as the Lepuropetalaceae.

Genera included: ***Lepuropetalon, Parnassia.***

76c (91) Hydrangeaceae. Herbs/soft-wooded shrubs, rarely climbing, many with stellate hairs. Leaves usually opposite, simple, exstipulate. Inflorescences various. Flowers mostly bisexual (sometimes the outer flowers of the inflorescence sterile and with enlarged corollas), actinomorphic (fertile flowers only). KCA hypogynous/perigynous/epigynous. K4–5, C4–7, A4–*n*, G(2–7)/rarely 1-celled; ovules *n*, axile/parietal; style 1, stigmas head-like/2–7, free or almost so. Capsule/berry.

There are about 17 genera and 170 species, several of the genera native to either Europe or North America. Many are cultivated as ornamental shrubs, especially *Deutzia, Hydrangea* and *Philadelphus*. *Philadelphus* and some of its allies (shrubs with stellate hairs; filaments often toothed beside the anther) are sometimes separated off as the family Philadelphaceae.

Genera included: ***Broussaisia, Cardiandra, Carpenteria, Decumaria, Deinanthe, Deutzia, Dichroa, Fendlera, Fendlerella, Hydrangea, Jamesia, Kirengeshoma, Philadelphus, Platycrater, Schizophragma, Whipplea.***

76d (92) Escalloniaceae. Trees/shrubs. Leaves mostly alternate, evergreen, exstipulate, usually with gland-tipped teeth. Flowers in racemes, actinomorphic, bisexual. KCA perigynous/epigynous. K(4–6) usually (5), C4–6 usually 5, A 4–6 usually 5, G usually (2); ovules *n*, parietal; style 1, somewhat bilobed at the apex. Capsule/berry. *Mainly southern hemisphere.*

There are about 15 genera with about 70 species. Some species and hybrids of the largest genus, *Escallonia*, are cultivated as ornamental shrubs and hedging plants. The family is often further divided (Iteaceae, Brexiaceae, etc.), but there is little agreement as to how this should best be done.

Genera included: ***Abrophyllum, Anopterus, Brexia, Carpodetus, Corokia, Escallonia, Itea, Quintinia***.

76e (93) Grossulariaceae. Shrubs, often spiny. Leaves alternate, simple/lobed, stipulate/exstipulate, usualy deciduous. Flowers in racemes, unisexual/bisexual, actinomorphic. KCA epigynous. K4–5, C4–5, A4–5, G(2); ovules few–*n*, parietal; styles free/united into a single style lobed at the apex. Berry. *Temperate northern hemisphere, South America.*

A single genus with about 150 species, native to both Europe and North America. Many species are cultivated as ornamental shrubs, others for their edible fruit (blackcurrant, redcurrant, gooseberry).

Genera included: ***Ribes***.

77 (94) Cunoniaceae. Woody. Leaves evergreen, opposite/whorled, trifoliolate or usually pinnate, usually stipulate; ptyxis conduplicate or supervolute. Flowers in racemes, heads, unisexual/bisexual, actinomorphic. KCA perigynous/epigynous. K4–5/4–10, C4–5/4–10/rarely 0, A4–10, anthers sometimes opening by pore-like slits, G(2–3); ovules *n*, axile; styles 2–3. Capsule/rarely drupe/nut. *Mostly southern hemisphere.*

There are 25 genera with about 350 species. Species of about 8 genera are occasionally cultivated. *Bauera*, which is from Australia and has trifoliate, exstipulate leaves and floral parts up to 10, is sometimes separated off as the family Baueraceae.

Genera included: ***Bauera, Caldcluvia, Callicoma, Cunonia, Geissois, Weinmannia***.

77a (95) Davidsoniaceae. Trees with irritant hairs. Leaves alternate, pinnate, large, with conspicuous, broad, palmately veined stipules. Flowers in racemes or panicles, bisexual, actinomorphic. PA hypogynous. P(5), A10 arising from a disc, G(2); ovules several per cell, axile. Fruit a drupe. *Australia.*

A family of a single species and genus (*Davidsonia pruriens*), very occasionally grown in European gardens.

Genera included: ***Davidsonia.***

78 (96) Pittosporaceae. Woody, sometimes climbing. Leaves alternate/opposite, simple, exstipulate, often evergreen; ptyxis variable. Inflorescence clusters/flowers solitary. Flowers bisexual, actinomorphic. KCA hypogynous. K5, C5, A5 anthers sometimes united, G(2–5); ovules *n*, axile; style 1, short, stigmas 2–5. Capsule/berry. *Tropics and south temperate Old World.*

There are 9 genera and about 240 species. Species of about 4 genera are cultivated as ornamental shrubs, especially *Pittosporum tobira.*

Genera included: ***Billardiera, Bursaria, Hymenosporum, Pittosporum, Sollya.***

79 (97) Byblidaceae. Herbs with insect-trapping glandular hairs on leaves. Leaves alternate, linear, spirally coiled when young, exstipulate. Flowers solitary, bisexual, actinomorphic. KCA hypogynous. K5, C5, A5 anthers opening by pores, G(2); ovules *n*, axile. Capsule. *N and W Australia.*

A single genus with 2 species, increasingly cultivated by insectivorous-plant enthusiasts.

Genera included: ***Byblis.***

80 (98) Roridulaceae. Small shrubs. Leaves with insect-trapping glandular hairs, alternate but clustered at the ends of branches, exstipulate, entire/lobed. Flowers solitary/in racemes, bisexual, actinomorphic. KCA hypogynous. K5, C5, A5 anthers opening by pores, G(3); ovules 1–4 per cell, axile; style 1, stigma capitate. Capsule. *South Africa.*

Another monogeneric family, sometimes included in Byblidaceae; like *Byblis*, increasingly found in cultivation in Europe.

Genera included: ***Roridula***.

81 (99) Bruniaceae. Low heather-like shrublets. Leaves alternate, needle-like, exstipulate. Flowers in spikes/heads, bisexual, actinomorphic. KCA hypogynous/epigynous. K4–5/(4–5), C4–5, A4–5, G(2–3); ovules 1–2 per cell, axile; styles 2–3, joined at extreme base. *South Africa.*

A family of 11 genera and 69 species. A few species of *Brunia* have, at times, been cultivated for the cut-flower trade.

Genera included: ***Brunia***.

82 (100) Rosaceae. Herbs/shrubs/trees/rarely climbing. Leaves usually alternate, rarely opposite/whorled, usually stipulate, simple/divided, evergreen/deciduous; ptyxis very variable (important in the classification of species of *Prunus*). Inflorescences various. Flowers usually actinomorphic and bisexual. KCA/PA perigynous (perigynous zone sometimes very small)/epigynous. K4–6/(4–6), epicalyx sometimes present, C4–6/rarely 0, A4–*n*, G1–*n*/(2–5); ovules 1–*n* per cell, axile/marginal; styles as many as the carpels, free or almost so/united for more than half their length. Fruit variable, follicles/achenes/'berries' of drupelets/pome. *Cosmopolitan.*

A large and variable family with 115 genera and about 3200 species. Many genera are native to Europe and/or North America, and very many are cultivated as ornamental herbs, shrubs or trees. Many produce edible fruit, notably *Rubus* (blackberry, raspberry), *Malus* (apple), *Pyrus* (pear) and *Prunus* (cherry, plum, almond, apricot, peach, nectarine). Includes Quillajaceae.

Genera included: ***Acaena, Adenostoma, Agrimonia, Alchemilla, Amelanchier, Aphanes, Aremonia, Aruncus, Bencomia, Cercocarpus, Chaenomeles, Chamaebatia, Chamaebatiaria, Chamaerhodos, Coleogyne, Coluria, Cotoneaster, Cowania, +Crataegomespilus, Crataegus, ×Cratae-Mespilus, Cydonia, Dichotomanthes, Docynia, Dryas, Duchesnea, Eriobotrya, Exochorda, Fallugia, Filipendula, Fragaria, Geum, Gillenia, Hagenia, Heteromeles, Holodiscus, Horkelia, Ivesia, Kageneckia, Kelseya, Kerria, Luetkea, Lyonothamnus, Maddenia, Malus, Margyricarpus, Mespilus, Neillia, Neviusia, Oemleria, Osteomeles, Peraphyllum, Petrophytum, Photinia, Physocarpus, Polylepis, Potentilla, Prinsepia, Prunus, Purpusia, Purshia, Pyracantha, Pyrus, Quillaja, Rhaphiolepis, Rhodotypos, Rosa, Rubus, Sanguisorba, Sarcopoterium, Sibbaldia, Sibiraea, Sorbaria, ×Sorbaronia, ×Sorbopyrus, Sorbus, Spiraea, Stephanandra, Vauquelinia, Waldsteinia.***

83 (101) Chrysobalanaceae. Woody. Leaves alternate, stipulate, often leathery. Flowers in racemes/panicles/cymes, usually bisexual, actinomorphic/zygomorphic. KCA perigynous. K5, C4–5, A3–*n*, G(2–3), often 1 or 2 carpels sterile, often asymmetrically placed in the tubular or cup-shaped perigynous zone; ovules 2, basal; styles free. Drupe. *Mainly tropics.*

There are 500 species in 17 genera. A few species are occasionally cultivated as glasshouse plants.

Genera included: ***Atuna, Chrysobalanus, Hirtella, Licania, Maranthes, Parinari.***

84 (102) Fabaceae (Leguminosae-Papilionoideae). Herbs/woody/climbing. Leaves usually alternate, stipulate, pinnately compound, sometimes 2–3-pinnate or of a single or 2–3 leaflets, rarely absent; ptyxis of leaflets almost aways conduplicate (but supervolute in some species of *Lathyrus*). Inflorescence various, usually a raceme. Flowers usually bisexual and zygomorphic. KCA hypogynous/perigynous. K usually 5/(5), C5, all free or sometimes the lower 2 united towards the apex, imbricate in bud such that the uppermost petal is outermost, overlapping the others, and often larger than them; A usually (10)/(9)+1, anthers opening by slits. G1/rarely more; ovules 1–*n*, marginal; style usually 1. Legume (sometimes indehiscent or breaking into 1-seeded segments); seeds usually without a lateral line. *Cosmopolitan.*

There are about 450 genera, and over 11 000 species. Many genera are native to Europe and/or North America, and many are grown as ornamentals. Many of the genera produce crops of economic importance (peas, beans, etc.). This and the following two families are often treated together as Leguminosae.

Genera included: ***Abrus, Adenocarpus, Adesmia, Aeschynomene, Alhagi, Alysicarpus, Amicia, Amorpha, Amphicarpaea, Anagyris, Andira, Anthyllis, Apios, Arachis, Argyrolobium, Astragalus, Baptisia, Biserrula, Bituminaria, Bolusanthus, Bossiaea, Brachysema, Brongniartia, Burtonia, Butea, Cajanus, Calicotome, Calophaca, Calopogonium, Campylotropis, Canavalia, Caragana, Carmichaelia, Castanospermum, Centrosema, Chamaecytisus, Chapmannia, Chordospartium, Chorizema, Christia, Cicer, Cladrastis, Clianthus, Clitoria, Cologania, Colutea, Coronilla, Crotalaria, Cytisus, Daviesia, Derris, Desmodium, Dillwynia, Dioclea, Diphysa, Dolichos, Ebenus, Echinospartium, Eremosparton, Erinacea, Eriosema, Errazurizia, Erythrina, Eversmannia, Eysenhardtia, Galactia, Galega, Genista, Gliricidia, Glycine, Glycyrrhiza, Gompholobium, Gonocytisus, Halimodendron, Hammatolobium, Hardenbergia, Hedysarum, Hippocrepis, Hovea, Hymenocarpos, Hypocalyptus, Indigofera, Kennedia, Laburnum, Lathyrus, Lembotropis, Lens, Lespedeza, Lonchocarpus, Lotononis, Lotus, Lupinus, Maackia, Machaerium, Macroptilium, Marina, Medicago, Melilotus, Mucuna, Mundulea, Myrospermum, Myroxylon, Neorudolphia, Nissolia, Notospartium, Olneya, Onobrychis,***

Ononis, Ormosia, Ornithopus, Oxylobium, Oxyrhynchus, Oxytropis, Pachyrhizus, Parochetus, Peteria, Petteria, Phaseolus, Pickeringia, Piptanthus, Piscidia, Pisum, Podalyria, Podocytisus, Pongamia, Psoralea, Psorothamnus, Pterocarpus, Pueraria, Pultenaea, Retama, Rhynchosia, Robinia, Sabinea, Scorpiurus, Sesbania, Sophora, Spartium, Sphaerolobium, Sphaerophysa, Sphinctospermum, Stauracanthus, Strongylodon, Strophostyles, Stylosanthes, Sutherlandia, Swainsona, Templetonia, Tephrosia, Teramnus, Thermopsis, Trifolium, Trigonella, Ulex, Vicia, Vigna, Viminaria, Virgilia, Wisteria, Zornia.

84a (103) Caesalpiniaceae (Leguminosae-Caesalpinioideae). Trees/shrubs/herbs. Leaves usually pinnate, rarely bipinnate, never reduced to phyllodes, stipulate. Flowers usually bisexual and zygomorphic. KCA hypogynous/perigynous. K usually 5/(5), C3–5, petals imbricate so that the uppermost is within the laterals, which themselves are overlapped by the 2 lower; A3–10, often deflexed downwards, anthers often opening by pores, pollen granular, G1; ovules 1–n, marginal. Legume; seeds usually without a lateral line. *Mainly tropics.*

There are 162 genera and about 2000 species. A few genera are native to Europe and/or North America, and species of various genera are grown as ornamentals, especially *Cercis* and *Gleditsia*, which are hardy in Europe, and *Delonix* and *Amherstia*, which are grown in warmer regions.

Genera included: ***Amherstia, Bauhinia, Brownea, Caesalpinia, Cassia, Ceratonia, Cercis, Delonix, Gleditsia, Gymnocladus, Haematoxylum, Hoffmannseggia, Hymenaea, Parkinsonia, Peltophorum, Schotia, Stahlia, Tamarindus.***

84b (104) Mimosaceae (Leguminosae-Mimosoideae). Mostly shrubs or trees, rarely herbaceous. Leaves usually bipinnate, rarely tripinnate or reduced to expanded stalks and rachises (phyllodes). Flowers in racemes or heads. KCA hypogynous. K often 4/(4), C 4/(4), actinomorphic, valvate in bud; A4–many, often conspicuous, anthers opening by slits, pollen sometimes in masses. G1; ovules 1–*n*, marginal. Legume; seeds with a U-shaped lateral line.

There are about 58 genera and 3100 species. Many are grown as ornamentals in Europe, especially species of *Acacia, Mimosa* and *Albizia*.

Genera included: ***Acacia, Adenanthera, Albizia, Anadenanthera, Calliandra, Desmanthus, Dichrostachys, Entada, Inga, Leucaena, Lysiloma, Mimosa, Neptunia, Pithecellobium, Prosopis, Schrankia***.

85 (105) Krameriaceae. Shrubs/herbs. Leaves alternate, entire, exstipulate. Inflorescence axillary/racemose. Flowers bisexual, zygomorphic. KCA hypogynous/K hypogynous, CA perigynous. K4–5, C5/(5), the lower pair often modified into glands, A3–4, anthers opening by pores; G1-celled; ovules 2, pendulous; style 1, stigma disc-like. Fruit 1-seeded, indehiscent, covered with barbed spines. *Mainly tropical America.*

A single genus with about 15 species, which occurs in the southern part of North America. One of the species is occasionally cultivated in Europe.

Genera included: ***Krameria***.

86 (106) Podostemaceae. Aquatics of running water, often resembling algae or mosses. Leaves alternate/rarely 0, simple. Flowers bisexual, zygomorphic. PA hypogynous. P2–3/(2–3), A1–4, G(2); ovules *n*, axile; styles 2, free, short. Capsule. *Mainly tropics*.

There are 40 genera with 275 species of very bizarre plants. They are mainly tropical, but 1 genus is native to North America.

Genera included: ***Podostemum***.

87 (107) Limnanthaceae. Herbs. Leaves alternate/basal, divided, exstipulate; ptyxis conduplicate. Flowers solitary, bisexual, actinomorphic. KCA hypogynous. K3–5, C3–5, A6–10, G(3–5), bodies of carpels free; style 1, divided above into as many stigmas as there are carpels; ovules 1 per carpel, ascending. Fruit a group of nutlets. *Temperate North America*.

Two genera with 6 species, native to western North America. Species of *Limnanthes* are widely cultivated.

Genera included: ***Floerkea, Limnanthes***.

88 (108) Oxalidaceae. Herbs/shrubs/trees. Leaves alternate/basal, pinnate/palmate/trifoliolate, exstipulate. Inflorescence various. Flowers bisexual, actinomorphic. KCA hypogynous. K5/(5), C5, contorted in bud, A10, filaments sometimes united into a tube, G(5); ovules 1–more, axile; styles 5, free. Capsule, often explosive. *Widespread*.

A family of 8 genera with 575 species. One genus (*Oxalis*) is native to both Europe and North America; many species of this genus are also in cultivation. *Averrhoa* is sometimes separated off as the Averrhoaceae.

Genera included: ***Averrhoa, Biophytum, Oxalis***.

89 (109) Geraniaceae. Usually herbs. Leaves alternate/opposite, simple/compound, stipulate/exstipulate; ptyxis conduplicate-plicate. Inflorescences various. Flowers bisexual/rarely unisexual, actinomorphic/zygomorphic. KCA usually hypogynous. K3–5/(3–5), sometimes the upper sepal spurred, the spur united with the flower-stalk, C2–5, A(5–15) often some infertile, G(3–5), carpel-bodies variously free, style long, beak-like, divided at the apex/3–5, free; ovules 1–*n* per cell, axile. Capsule/berry/schizocarp. *Widespread*.

There are 14 genera with 700 species. Species of *Geranium* are native to Europe and North America, and species of *Erodium* to

Europe. Many species are cultivated, especially from these genera and *Pelargonium*. Includes Biebersteiniaceae.

Genera included: ***Biebersteinia, Erodium, Geranium, Pelargonium, Sarcocaulon.***

90 (110) Tropaeolaceae. Herbs. Leaves alternate/opposite, simple/divided, stipulate/exstipulate; ptyxis flat or conduplicate. Flowers bisexual, zygomorphic, solitary, axillary, KC partly perigynous, A hypogynous. K5 spurred, C2/5, A8, G(3); ovules 1 per cell, axile; style 1, lobed at the apex. Schizocarp. *Central and South America.*

A single genus with about 80 species. Several are cultivated as ornamental climbers or scramblers.

Genera included: ***Tropaeolum.***

91 (111) Zygophyllaceae. Herbs/shrubs. Leaves usually opposite, usually compound, stipulate, often fleshy; ptyxis variable. Inflorescence cymose/flowers solitary. Flowers usually bisexual and actinomorphic. KCA hypogynous, disc usually present. K4–5, C4–5, A5–15, G(2–5); ovules *n*, axile; style 1/stigmas sessile. Capsule/drupe-like/schizocarp.

A rather variable family with 27 genera and 250 species, mostly desert plants, sometimes succulent. Six genera native to Europe, 7 to North America. Only a small number are cultivated. *Nitraria* is sometimes separated off as the Nitrariaceae, *Peganum* as the Peganaceae and *Tetradiclis* as the Tetradiclidaceae.

Genera included: ***Fagonia, Guaiacum, Kallstroemia, Larrea, Nitraria, Peganum, Porlieria, Tetradiclis, Tribulus, Zygophyllum.***

92 (112) Linaceae. Herbs/shrubs. Leaves alternate/opposite, entire, stipulate/exstipulate; ptyxis flat or conduplicate. Inflorescence a cyme. Flowers bisexual, actinomorphic. KCA hypgynous. K4–5/(4–5), C3–5, A4–5/10/15, sometimes united at base,

G(3–5), usually 6–10-celled by the growth of secondary septa; ovules 1–2 per cell, axile; styles 3–5, free. Capsule/drupe. *Widespread.*

A mainly tropical family of 15 genera and over 500 species. Two genera are native to Europe and 4 to North America. A few genera are cultivated for their flowers; flax and linseed are extracted from *Linum usitatissimum*. Some of the tropical members (not included here) are sometimes separated off as Hugoniaceae.

Genera included: ***Hesperilinon, Linum, Radiola, Reinwardtia, Sclerolinon.***

93 (113) Erythroxylaceae. Woody. Leaves alternate, simple, stipulate; ptyxis revolute. Inflorescences various. Flowers bisexual, actinomorphic. KCA hypogynous. K5, C5 each with an appendage on the inner face, A(10), filaments united at base, G(3), often only 1 cell developing; ovules 1–2 per cell, axile; styles free. Fruit berry-like. *Tropics (mainly America).*

A small, rather uniform family of 4 genera and 250 species. One genus is native to North America. *Erythroxylon coca*, the original source of the drug cocaine, is sometimes grown as a curiosity in European glasshouses.

Genera included: ***Aneulophus, Erythroxylon.***

94 (114) Euphorbiaceae. Woody/herbs/succulents, milky sap often present, frequently acrid and poisonous. Leaves usually alternate and stipulate, simple/compound, rarely absent, sometimes replaced by cladodes; ptyxis very variable. Inflorescences various, sometimes a cup of bracts with glandular margins (cyathium). Flowers unisexual, actinomorphic. PA/KCA hypogynous. P4–6/(2–6)/rarely K5/10, C5/(5) A1–*n*/(2–*n*), G(2–4)/ rarely more, usually (3); ovules 1–2 per cell, axile; styles 2–4/ rarely more, usually 3, free or slightly joined at the base, often divided above. Fruit usually schizocarpic; seeds often carunculate. *Widespread.*

A large and variable family of 326 genera and 7750 species. Seven genera are native to Europe, 47 to North America. Species of about 20 are cultivated for ornament; the most important is *Euphorbia*, which contains both normal herbaceous and shrubby species and succulent, more or less leafless shrubs. Includes Phyllanthaceae.

Genera included: ***Acalypha, Adelia, Alchornea, Alchorneopsis, Aleurites, Andrachne, Antidesma, Argythamnia, Bernardia, Bischofia, Breynia, Caperonia, Chrozophora, Claoxylon, Cnidoscolus, Codiaeum, Dalechampia, Ditaxis, Ditta, Drypetes, Eremocarpus, Euphorbia, Excoecaria, Flueggea, Garcia, Gymnanthes, Hevea, Hippomane, Homalanthus, Hura, Hyeronima, Jatropha, Julocroton, Macaranga, Mallotus, Manihot, Margaritaria, Mercurialis, Monadenium, Pedilanthus, Phyllanthus, Reverchonia, Ricinus, Sapium, Sauropus, Savia, Sebastiania, Securinega, Stillingia, Synadenium, Tetracoccus, Tragia.***

95 (115) Daphniphyllaceae. Trees/shrubs. Leaves alternate, crowded, entire, exstipulate, usually evergreen; ptyxis flat. Flowers in axillary racemes, unisexual, actinomorphic. PA hypogynous. Male flowers P3–8, imbricate, A6–12; female flowers P0, staminodes few, small/0, G(2), imperfectly divided; styles 1–2, persistent, undivided; ovules 2 per cell, pendulous. Drupe, 1-seeded. *Temperate eastern Asia.*

A small family with a single genus and about 10 species, 1 or 2 of which are grown for ornament.

Genera included: ***Daphniphyllum.***

96 (116) Rutaceae. Woody/herbaceous. Leaves alternate/opposite, simple/compound, exstipulate, usually aromatic, gland-dotted, often evergreen; ptyxis usually conduplicate, rarely flat. Inflorescences various. Flowers usually bisexual, usually actinomorphic. KCA usually hypogynous, disc usually present. K3–6/(3–6), C3–6/(3–6)/rarely 0, A3–12/rarely more, staminodes sometimes present, G(4–5)/rarely (*n*), carpels often free but united by single styles/otherwise styles rarely free, as many as carpels, ovary often borne on a short stalk; ovules 1–*n*, axile. Fruit fleshy/capsule/samaras. *Tropics, warm temperate areas.*

A highly variable family of 160 genera and 1650 species, with 6 genera occurring in Europe (3 introduced) and 20 native to North America. Most species in the family have characteristic glands in the leaves, which contain aromatic oils, giving rise to identifiable scents when crushed. Species of many genera are cultivated for ornament and the citrus fruits (oranges, lemons, grapefruit, limes, etc.) belong to *Citrus* and its allies.

Genera included: ***Acradenia, Adenandra, Aegle, Amyris, Boenninghausenia, Boronia, Choisya, Citrus, Cneoridium, Coleonema, Correa, Crowea, Dictamnus, Eriostemon, Esenbeckia, Fortunella, Glycosmis, Haplophyllum, Helietta, Limonia, Melicope, Murraya, Orixa, Phellodendron, Pilocarpus, Platydesma, Poncirus, Ptelea, Ravenia, Ruta, Skimmia, Thamnosma, Triphasia, Zanthoxylum, Zieria.***

97 (117) Cneoraceae. Shrubs, sometimes with medifixed hairs. Leaves alternate, simple, exstipulate; ptyxis conduplicate. Inflorescence a cyme. Flowers bisexual, actinomorphic. KCA hypogynous, disc 0. K3–4, C3–4, A3–4, G(3–4), stalked; ovules 1–2 per cell, axile; styles 3–4, fre. Schizocarp. *Mediterranean area, Canary Islands, Cuba?*

A small family of a single genus, which is occasionally further divided, with 3 species (1 native to Europe and often cultivated). The distribution of the family is unusual and striking.

Genera included: ***Cneorum.***

98 (118) Simaroubaceae. Woody. Leaves alternate, simple/compound, exstipulate; ptyxis conduplicate. Inflorescences various. Flowers usually unisexual, actinomorphic. KCA hypogynous, disc present. K(3–8), C0–8, A4–14/rarely more, G(2–5);

ovules 2 per cell, axile; styles 1–5, free. Fruits various, often samaras. *Mainly tropics.*

A diverse family of 23 genera and about 170 species. Nine genera are native to North America, and one is introduced in Europe. A few are cultivated as ornamental trees. *Kirkia* is sometimes separated off as the Kirkiaceae, and *Picramnia* as the Picramniaceae.

Genera included: ***Ailanthus, Alvaradoa, Castela, Kirkia, Picramnia, Picrasma, Quassia, Simarouba.***

99 (119) Burseraceae. Woody with aromatic resins. Leaves alternate, usually compound, exstipulate. Inflorescences panicles/flowers solitary. Flowers unisexual/bisexual, actinomorphic. KCA hypogynous with a disc. K(3–5), C3–5/rarely 0, A6–10, G(2–5); ovules 2 per cell, axile; style 1 or stigma sessile. Drupe/capsule. *Mainly tropics.*

There are 18 genera with 540 species. Three genera are native to North America. A few species are cultivated as aromatic shrubs.

Genera included: ***Bursera, Canarium, Dacryodes, Tetragastris.***

100 (120) Meliaceae. Trees, wood often scented. Leaves usually alternate, mostly pinnate, exstipulate; ptyxis conduplicate. Inflorescences cymose panicles. Flowers usually bisexual, actinomorphic/rarely slightly zygomorphic. KCA hypogynous with disc. K(4–6), C4–6/rarely 8, A(8–12)/rarely free/fewer/more, G(2–20); ovules usually 2 or more per cell, axile; style 1, short. Berry/capsule/drupe. *Mainly tropics.*

A family of 51 genera and about 570 species. Five genera are native to North America and 1 is introduced into Europe. Species of about 7 genera are grown as ornamental shrubs in glasshouses, and as street trees in southern Europe.

Genera included: ***Azadirachita, Cedrela, Guarea, Melia, Swietenia, Toona, Trichilia, Turraea.***

101 (121) Malpighiaceae. Woody, often with medifixed hairs. Leaves usually opposite/rarely alternate/whorled, simple/rarely lobed/divided, stipulate/exstipulate, usually evergreen; ptyxis flat or conduplicate. Inflorescences racemes/umbels/panicles. Flowers usually bisexual, slightly zygomorphic, rarely actinomorphic. KCA hypogynous. K5, often some or all sepals with 2 nectar-secreting glands on the outside, C5 often fringed and/or petals unequal, A(10) anthers sometimes opening by pores, sometimes some sterile; G(3); ovules 1 per cell, axile; styles 3, free/united. Fruits various, chiefly winged mericarps. *Mainly tropical America.*

Sixty-six genera and about 1100 species, with 11 genera native to North America. Species of a few genera are cultivated in Europe.

Genera included: ***Acridocarpus, Aspicarpa, Banisteriopsis, Bunchosia, Byrsonima, Galphimia, Heteropterys, Hiptage, Janusia, Malpighia, Sphedamnocarpus, Stigmaphyllon, Tetrapterys, Tristellateia.***

102 (122) Tremandraceae. Shrublets. Leaves usually opposite, entire, exstipulate. Flowers solitary, bisexual, actinomorphic. KCA hypogynous. K4–5, C4–5, A8–10, anthers opening by pores, G(2); ovules 1–2/rarely 3 per cell, axile; style 1, short, stigma terminal. Capsule. *Australia.*

There are 3 genera with about 46 species. A few species have been cultivated in Europe from time to time, but they do not persist.

Genera included: ***Platytheca, Tetratheca.***

103 (123) Polygalaceae. Herbs/shrubs. Leaves usually alternate, entire, exstipulate; ptyxis flat or supervolute. Inflorescences racemes/spikes/panicles. Flowers bisexual, zygomorphic. KCA hypogynous/K hypogynous, CA perigynous. K usually 5, lateral pair petal-like, C usually 3, often joined to staminal tube, A(8–10)/rarely (4–5), anthers opening by pores, G usually (2);

ovules usually 1 per cell, axile; style 1, stigmas as many as carpels. Fruit usually a capsule; seeds with arils. *Widespread.*

A mainly tropical family of 18 genera and 950 species. One genus is native to Europe and 3 to North America. A few species of *Polygala* are found in cultivation.

Genera included: ***Monnina, Nylandtia, Polygala, Securidaca.***

104 (124) Coriariaceae. Shrubs, branches angular. Leaves opposite, entire, exstipulate, ptyxis flat. Flowers solitary or in racemes, usually bisexual, actinomorphic. KCA hypogynous. K5, C5 keeled inside, A10, G5–10; ovules 1 per carpel, apical; styles as many as carpels. Achene surrounded by fleshy C. *Scattered.*

A single genus with perhaps 12 species, native to Europe and occasionally found in cultivation.

Genera included: ***Coriaria.***

105 (125) Anacardiaceae. Woody with resinous bark. Leaves alternate, simple/compound, exstipulate; ptyxis conduplicate or rarely flat. Inflorescences racemes/panicles. Flowers bisexual/unisexual, actinomorphic. KCA hypogynous, disc often present. K3–5/(5), C3–5 rarely 0, A1–10, G(3–5); ovules 1 per cell, apical/basal; style always 1, sometimes divided above. Drupe, 1-seeded. *Tropics and warm temperate areas.*

A mainly tropical family of resinous trees and shrubs with 73 genera and 850 species. Four genera (1 introduced) occur in Europe and 11 in North America. Species of about 10 genera are cultivated in Europe. *Mangifera indica* produces the mango and *Anacardium occidentale* the cashew nut.

Genera included: ***Anacardium, Comocladia, Cotinus, Harpephyllum, Loxostylis, Mangifera, Metopium, Pistacia, Rhus, Schinopsis, Schinus, Sclerocarya, Semecarpus, Spondias.***

106 (126) Aceraceae. Woody, sap sometimes milky. Leaves usually opposite, simple/compound (usually palmately lobed), exstipulate; ptyxis conduplicate-plicate. Flowers clustered/in racemes, mostly functionally unisexual (appearing bisexual), actinomorphic. KC perigynous, A hypogynous/perigynous, disc present. K4–5, C4–5/rarely 0, A4–5, G(2–3); ovules 2 per cell, axile; styles joined in their lower half. Winged mericarps. *North temperate areas.*

There are 2 genera with over 100 species. One genus is native to both Europe and North America, and species of both are cultivated.

Genera included: ***Acer, Dipteronia.***

107 (127) Sapindaceae. Trees/shrubs/woody climbers with tendrils in the inflorescences. Leaves usually alternate and compound, stipulate/exstipulate; ptyxis conduplicate. Inflorescences cymose/racemose/panicles. Flowers unisexual (often superficially bisexual), actinomorphic/zygomorphic. KCA hypogynous, disc outside A, often elaborate. K4–5 often unequal, C4–5/rarely 0, A5–8, G usually (3); ovules 2 per cell, axile; style 1, short, sometimes lobed above. Fruits various; seeds with arils. *Mostly tropics.*

One hundred and forty-five genera and about 1300 species. One genus occurs in Europe (introduced) and 17 are native to North America. Species of about 8 genera are in cultivation in Europe.

Genera included: ***Alectryon, Allophylus, Cardiospermum, Dodonaea, Exothea, Filicium, Hypelate, Koelreuteria, Matayba, Melicoccus, Paullinia, Sapindus, Serjania, Talisia, Thouinia, Ungnadia, Urvillea, Xanthoceras.***

108 (128) Hippocastanaceae. Woody. Leaves opposite, palmate, exstipulate; ptyxis conduplicate. Inflorescence racemose. Flowers usually bisexual, zygomorphic. KC perigynous, A usually hypogynous, disc present. K(4–5), C4–5, A5–9, G(3); ovules 2 per cell, axile; style 1. Capsule containing large seeds. *Tropics and temperate areas.*

There are 2 genera, one native to both Europe and North America, with about 15 species. Several species and hybrids are cultivated.

Genera included: ***Aesculus***.

109 (129) Sabiaceae. Woody. sometimes climbing. Leaves alternate, evergreen/deciduous, simple, exstipulate. Panicles. Flowers bisexual/unisexual, actinomorphic. KCA hypogynous. K5, C5, A usually 5 and on the same radii as the petals, G(2), carpels only slightly united; style 1; ovules axile, 2 per cell. Fruit a pair of drupes. *Himalaya, south and south-east Asia.*

A single genus with about 30 species, a few of them occasionally cultivated.

Genera included: ***Sabia***.

109a (130) Meliosmaceae. Woody, evergreen. Leaves alternate, pinnate, exstipulate. Flowers in panicles, bisexual/unisexual, zygomorphic. KCA hypogynous. K5, 3 large + 2 smaller, A2 fertile + 3 staminodes, all on the same radii as the petals, G(2–3); style 1, ovules axile, 1 per cell. *Southeast Asia, tropical America.*

Two genera and perhaps 20 species; a few species of *Meliosma* are cultivated as interesting small trees.

Genera included: ***Meliosma***.

110 (131) Melianthaceae. Herbaceous/woody. Leaves alternate, pinnate, stipulate, stipules between leaf-stalk and stem, often large; ptyxis conduplicate or conduplicate-plicate. Racemes. Flowers unisexual/bisexual, zygomorphic. KC perigynous, A hypogynous, with disc. K4–5, C4–5, A4–5/rarely 10, free/united, G(4–5); ovules 1–*n* per cell, axil; style 1, stigma 4–5-lobed. Capsule. *Africa.*

Two genera with about 15 species.

Genera included: ***Bersama, Melianthus***.

110a (132) Greyiaceae. Soft-wooded trees. Leaves alternate, simple, exstipulate. Flowers in racemes, bisexual, actinomorphic. KCA perigynous, disc present. K5, persistent, C5, A10, borne on the disc, which itself bears 10 staminode-like projections, G(5), 1-celled but placentas very intrusive; style 1; ovules *n*, parietal. Capsule. *Southern Africa.*

A single genus with 3 species, 1 of which is quite frequently cultivated.

Genera included: ***Greyia.***

111 (133) Balsaminaceae. Herbs. Leaves alternate, whorled, simple, exstipulate; ptyxis involute. Flowers solitary/clusters/ racemes (occasionally umbel-like). Flowers bisexual, zygomorphic. KCA hypogynous. K3/5, often coloured, the lowest spurred, C5 (the lateral sometimes fused in 2 pairs), A(5), the anthers cohering above the ovary like a cap, G(5); ovules *n*, axile; style 1, very short. Explosive capsule. *Mostly Old World.*

A family of 2 genera with almost 1000 species; one genus native to both Europe and North America.

Genera included: ***Impatiens.***

112 (134) Cyrillaceae. Woody. Leaves alternate, simple, exstipulate. Inflorescences racemose. Flowers bisexual, actinomorphic. KCA hypogynous, disc 0. K(5), C5/(5), A5/10, G(2–4); ovules 1–2 per cell, axile; style 1, short, stigmas 2–4. Fruit dry, indehiscent. *Mainly tropical America.*

A small family of 3 genera and 14 species; 2 of the genera are native to the southern part of North America, and 1 is occasionally cultivated in Europe.

Genera included: ***Cliftonia, Cyrilla.***

113 (135) Aquifoliaceae. Woody. Leaves alternate, simple, often evergreen, stipules small and falling early; ptyxis supervolute. Flowers in cymes/clusters, functionally unisexual/

rarely bisexual, actinomorphic. KCA hypogynous, disc 0. K4–5, C4–5/(4–5), A4–5, sometimes attached to base of C, G(3–*n*); ovules 1–2 per cell, axile; style 0, stigmas sessile. Drupe. *Widespread.*

Two genera (sometimes treated as one), with about 400 species. Both genera are native to North America, 1 to Europe; both are widely cultivated.

Genera included: ***Ilex, Nemopanthus.***

114 (136) Corynocarpaceae. Woody. Leaves alternate, stipulate, evergreen; ptyxis conduplicate. Flowers in panicles, bisexual, actinomorphic. KCA hypgynous. K5, C5, A5 on same radii as petals and attached to them at the extreme base, G(2); ovule 1, pendulous; styles 1 or 2. Drupe. *Asutralasia, Polynesia.*

A single genus with 4 species, occasionally seen in cultivation.

Genera included: ***Corynocarpus.***

115 (137) Celastraceae. Woody, sometimes climbing. Leaves alternate/opposite, simple, stipulate/exstipulate; ptyxis mostly involute. Inflorescences racemes/panicles, rarely flowers solitary. Flowers unisexual/bisexual, actinomorphic. KCA hypogynous/perigynous, with conspicuous disc. K4–5, C4–5, A4–5, G(2–5); ovules usually 2 per cell, axile; styles 1 or as many as ovary cells. Fruit a capsule/berry/drupe/samara, seeds with arils. *Widespread.*

A rather uniform, mainly tropical family with 70 genera and perhaps 1300 species. Two genera are native to Europe and 13 to North America; species of about 7 are cultivated as ornamental trees and shrubs. *Hippocratea* is sometimes separated off as the Hippocrateaceae.

Genera included: ***Cassine, Catha, Celastrus, Euonymus, Gyminda, Hippocratea, Maytenus, Mortonia, Paxistima, Perrottetia, Ptelidium, Putterlickia, Schaefferia, Torralbasia, Tripterygium.***

115a (138) Canotiaceae. Small trees or shrubs, branches ending in spines. Leaves reduced to small, alternate scales, exstipulate. Flowers in cymes, bisexual, actinomorphic, with disc. KCA hypogynous. K(5), C5, A5, G(5); ovules up to 6 per carpel, axile; style 1. Capsule. *Southwest USA.*

A family of a single genus and species, native to North America.

Genera included: ***Canotia***.

116 (139) Staphyleaceae. Trees/shrubs. Leaves usually opposite, compound/rarely simple, stipulate (stipules falling early); ptyxis involute. Inflorescence of drooping panicles/racemes. KCA perigynous, with disc. K(5), C5, A5, G(2–4); ovules *n*, axile; styles 2–4, free. Inflated capsule. *North temperate areas, South America, Asia.*

There are 5 genera and about 60 species. One genus is native to Europe, 2 to North America. A few species of *Staphylea* are grown as flowering shrubs with curious, inflated and bladdery fruits.

Genera included: ***Staphylea, Turpinia***.

117 (140) Stackhousiaceae. Herbs with rhizomes. Leaves alternate, simple, exstipulate. Inflorescences spikes/racemes/clusters. Flowers bisexual, actinomorphic. KCA perigynous. K(5), C5, lobes either entirely free or free at base and apex, sometimes united in the middle, A5, G(2–5); ovules 1 per cell, axile/basal; style 1, divided into 2–5 stigmas at about half its length. Schizocarp. *Australasia.*

A uniform family of 3 genera and 38 species, 1 or 2 of which are occasionally cultivated in Europe.

Genera included: ***Stackhousia***.

118 (141) Buxaceae. Evergreen, usually woody. Leaves alternate/opposite, simple, exstipulate. Flowers usually unisexual, actinomorphic. PA hypogynous. P4–6, A4, G(2–3); ovules 1–2 per cell, axile; styles free. Capsule/berry-like, seeds shiny black. *Tropical and temperate areas.*

A family of 4 genera and about 60 species; one genus native to Europe, 3 to North America. Species of all genera are cultivated, and *Buxus* is grown for its hard, smooth wood.

Genera included: ***Buxus, Pachysandra, Sarcococca.***

118a (142) Simmondsiaceae. Rigid, evergreen, dioecious shrubs. Leaves opposite, rigid, simple, exstipulate. Male flowers in heads, PA hypogynous; female flowers solitary. P4–6, A8–12, G(3); styles 3, free; ovules 1 per cell, axile. Capsule. *Western USA.*

A single genus and species, cultivated for the production of jojoba oil.

Genera included: ***Simmondsia.***

119 (143) Icacinaceae. Trees/shrubs/climbers. Leaves alternate, simple, exstipulate; ptyxis conduplicate. Inflorescence a panicle. Flowers unisexual/bisexual, actinomorphic. KCA hypogynous. K(4–5), C4–5, A5, G(2–3); ovules 2 per cell, axile, pendulous; style 1, short, divided above. Drupe/samaras. *Mainly tropics.*

There are about 60 genera with 320 species; two of the genera are native to North America, and species of a few are cultivated in Europe.

Genera included: ***Apodytes, Mappia, Ottoschulzia, Pennantia.***

120 (144) Rhamnaceae. Woody. Leaves usually alternate and stipulate, simple; ptyxis conduplicate or involute. Inflorescences corymbs/cymes/clusters. Flowers unisexual/bisexual, actinomorphic. KCA perigynous/epigynous, disc usually present. K4–5, C4–5/rarely 0, A4–5 on same radii as petals (or on radii alternating with sepals if petals absent), G(2–4); ovules 1/rarely 2 per cell, axile; styles joined in the basal half. Capsule/drupe-like. *Tropics, north temperate areas.*

There are 53 genera and about 875 species. Four genera native to Europe, 14 to North America. Species of several genera are grown as ornamental shrubs and small trees.

Genera included: ***Adolphia, Alphitonia, Berchemia, Ceanothus, Collettia, Colubrina, Condalia, Discaria, Gouania, Hovenia, Karwinskia, Krugiodendron, Paliurus, Phylica, Pomaderris, Reynosia, Rhamnus, Sageretia, Ziziphus.***

121 (145) Vitaceae. Usually woody climbers with tendrils/ rarely shrubs/trees/succulents. Leaves alternate, simple/ compound, stipulate/exstipulate, ptyxis conduplicate. Flowers small, in cymes, unisexual/bisexual, actinomorphic. KCA perigynous, disc present. K4–6/(4–6), C4–6/(4–6), when united falling as a unit, A4–6, on the same radii as the petals, G(2–6); ovules 1–2 per cell, axile. Berry. *Tropical and warm temperate areas.*

A family mainly of climbers in 13 genera and about 800 species. Two genera (1 introduced) occur in Europe and 4 in North America. About 8 genera are cultivated for ornament, and *Vitis vinifera* is widely grown for grapes.

Genera included: ***Ampelopsis, Cayratia, Cissus, Cyphostemma, Parthenocissus, Rhoicissus, Tetrastigma, Vitis.***

122 (146) Leeaceae. Shrubs, rarely scrambling/rarely herbaceous. leaves alternate, simple/compound, exstipulate (though stalk-bases swollen). Flowers in cymes/panicles, bisexual, actinomorphic. KCA perigynous, with disc. K4–5/(4–5), C4–5, A4–5 on same radii as petals, G(4–8); ovules 1 per cell, axile; style 1. Berry. *Old World tropics.*

A single genus with 34 species, a few of which are cultivated in Europe.

Genera included: ***Leea***.

123 (147) Elaeocarpaceae. Woody. Leaves alternate/opposite/rarely whorled, stipulate, deciduous or evergreen; ptyxis variable. Flowers in cymes/racemes/panicles/clusters, unisexual/bisexual, actinomorphic. KCA hypogynous/perigynous, K4–5, C1–5/0, A 4/*n*, anthers opening by short pore-like slits or by full-length slits, G(2–5); ovules *n*, axile; style 1, lobed at apex. Capsule/drupe-like. *Mainly southern hemisphere*.

A family of 11 genera and 220 species. Three genera occur in North America and species of 4 are cultivated as ornamental trees and shrubs.

Genera included: ***Aristotelia, Crinodendron, Elaeocarpus, Sloanea, Vallea***.

124 (148) Tiliaceae. Trees/shrubs/rarely herbs, often with stellate hairs. Leaves alternate, simple, mucilaginous, evergreen/deciduous, stipulate (stipules often falling early); ptyxis conduplicate or supervolute. Flowers in cymes, bisexual, actinomorphic. KCA hypogynous, K4–5/(4–5), C4–5, sometimes with nectaries on the petal-claws, A*n*, sometimes in 5 bundles and/or the outermost sterile, G(3–5); ovules 1–*n* per cell, axile; style 1, very shortly lobed. Capsule/drupe/indehiscent. *Widespread*.

There are 48 genera and 725 species. One genus is native to Europe, 3 to North America. Species of about 5 genera are grown as ornamental trees and shrubs, especially *Tilia*, which is used as a street tree or for forming avenues. *Muntingia* is sometimes separated off as Muntingiaceae and *Sparrmannia* as the Sparrmanniaceae.

Genera included: ***Corchorus, Entelea, Grewia, Muntingia, Prockia, Sparrmannia, Tilia, Triumfetta***.

125 (149) Malvaceae. Herbs/woody, often with stellate hairs. Leaves alternate, simple/divided, softly mucilaginous, stipulate; ptyxis variable. Inflorescences various. Flowers usually bisexual, actinomorphic. K hypogynous, CA perigynous. K5/(5), often with epicalyx, often with nectary-patches on the inner surface, C5, contorted, free from each other but all united at the base to the staminal tube, A(5–*n*), anthers 1-celled, G(2–*n*); ovules 1–*n* per cell, axile; style 1 divided in upper third or less. Capsule/schizocarp/berry. *Widespread.*

A large and usually easily recognised family of 121 genera and about 1550 species. Fourteen genera (4 of them introduced) occur in Europe and 41 in North America. Species of about 25 genera are cultivated for ornament. Species of *Gossypium* provide the cotton of commerce. The next two families are often considered to form part of the Malvaceae.

Genera included: ***Abelomoschus, Abutilon, Alcea, Allowissadula, Althaea, Anisodontea, Anoda, Bastardia, Batesimalva, Callirhoe, Cienfuegosia, Eremalche, Fryxellia, Gossypium, Herissantia, Hibiscadelphus, Hibiscus, Hoheria, Horsfordia, Iliamna, Kitaibela, Kokia, Kosteletzkya, Lagunaria, Lavatera, Malachra, Malacothamnus, Malope, Malva, Malvastrum, Malvaviscus, Malvella, Meximalva, Modiola, Napaea, Nototriche, Pavonia, Phymosia, Plagianthus, Sida, Sidalcea, Sphaeralcea, Thespesia, Urena, Wissadula.***

126 (150) Bombacaceae. Trees often with swollen trunks. Leaves simple/palmate, often scaly, stipules deciduous; ptyxis conduplicate, rarely conduplicate-plicate. Flowers large, bisexual, actinomorphic. KCA hypogynous/CA perigynous, K5/(5), C5 crumpled in bud, A5–*n*/(5–*n*), anthers 1-celled. G(2–5); ovules 2–*n* per cell, axile; style 1, stigmas capitate. Capsule/indehiscent; seeds often embedded in wool. *Mainly tropics.*

Thirty genera with about 250 species. There are 4 genera native to the southern parts of the USA, and species of a few are cultivated as ornamental glasshouse trees.

Genera included: ***Adansonia, Bombax, Ceiba, Chorisia, Durio, Ochroma, Pachira, Pseudobombax, Quararibea.***

127 (151) Sterculiaceae. Usually woody, often with stellate hairs. Leaves alternate, simple/compound, stipulate; ptyxis flat, conduplicate or conduplicate/plicate. Flowers solitary or clustered, unisexual/bisexual, actinomorphic. KCA hypogynous. K(5), C5/rarely 0, A(5/10), rarely free, G(2–5); ovules 2–*n* per cell, axile; styles 2–5, free/single, lobed at apex. Fruits various, sometimes splitting into apparently free carpels (follicles). *Mostly tropics.*

There are 73 genera and 1500 species. Eleven genera are native to North America, and species of 13 genera are cultivated. Includes Helicteraceae.

Genera included: ***Abroma, Ayenia, Brachychiton, Cola, Dombeya, Firmiana, Fremontodendron, Guazuma, Guichenotia, Helicteres, Hermannia, Melochia, Pentapetes, Reevesia, Sterculia, Theobroma, Triplochiton, Waltheria.***

128 (152) Thymelaeaceae. Usually woody. Leaves alternate/opposite, simple, exstipulate; ptyxis conduplicate/supervolute. Flowers in heads/racemes/spikes/rarely solitary, usually bisexual, actinomorphic, PA perigynous/rarely KCA perigynous. P(4–6), often tubular, coloured/rarely K4 C4, small, A2–8, G1; ovules 1–2, more or less apical; style 1. Drupe/nut/capsule. *Widespread.*

A family of about 50 genera and 750 species. Three genera are native to Europe, 6 to North America. Species of about 10 genera are cultivated as ornamental shrubs.

Genera included: ***Dais, Daphne, Diarthron, Dirca, Drapetes, Edgeworthia, Gnidia, Ovidia, Passerina, Pimelea, Stellera, Thymelaea.***

129 (153) Elaeagnaceae. Woody, with conspicuous silvery or brown scales on all parts. Leaves alternate/opposite, entire, deciduous/evergreen, exstipulate; ptyxis variable. Flowers solitary/in clusters/racemes/spikes, unisexual/bisexual, actinomorphic. PA perigynous. P(2–6), A4/8, G1; ovule 1, basal; style 1. Achene in persistent, fleshy P. *Widespread.*

There are 3 genera with 45 species. Two genera are native to Europe and 3 to North America. Species of all 3 genera are grown as shrubs with ornamental foliage.

Genera included: ***Elaeagnus, Hippophae, Shepherdia.***

130 (154) Flacourtiaceae. Shrubs/trees/woody climbers, sometimes spiny. Leaves alternate, simple, stipulate; ptyxis mainly supervolute. Inflorescence racemes/panicles/axillary clusters. Flowers unisexual/bisexual, actinomorphic. PA/KCA hypogynous/epigynous (in some non-cultivated genera). P4–5/3+6, A4–5/7–10/*n*, G(2–5); ovules *n*, parietal; style 1/styles free/stigmas sessile. Capsule/berry. *Mostly tropics.*

A large family of about 90 genera and 1250 species. Nine genera are native to North America and species of about 6 are in cultivation. *Berberidopsis* (2 species, climbers with perianth of 3+6 reddish segments) is sometimes separated off as Berberidopsidaceae.

Genera included: ***Azara, Berberidopsis, Carrierea, Casearia, Flacourtia, Idesia, Poliothyrsis, Xylosma.***

131 (155) Violaceae. Herbs/shrubs. Leaves alternate/basal, stipulate, simple/divided; ptyxis involute. Flowers solitary/in clusters, actinomorphic/zygomorphic, usually bisexual. KCA hypogynous. K5/(5), often persistent, C5, 1 often spurred, A5, connectives appendaged, G(2–3); ovules 1–*n*, parietal; style 1 or styles almost free, divided above, sometimes 2-, 3- or 6-fid. Capsule/berry. *Widespread.*

About 20 genera and almost 1000 species. *Viola* is native to Europe and it and 3 other genera are native to North America. *Viola* is very widely cultivated, as are a few species from other genera.

Genera included: ***Hybanthus, Isodendrion, Melicytus, Viola.***

132 (156) Stachyuraceae. Shrubs or small trees. Leaves alternate, simple, stipulate (stipules falling early); ptyxis involute. Inflorescence racemose/spicate. Flowers bisexual, actinomorphic. KCA hypogynous. K4, C4, A8, G(4); ovules *n*, axile; style 1. Berry. *East Asia.*

A single genus with 5 or 6 species, a few of which are grown as ornamental shrubs.

Genera included: ***Stachyurus.***

133 (157) Turneraceae. Shrubs/herbs. Leaves alternate, simple/lobed, exstipulate, often with 2 glands at base of blade. Flowers solitary/in clusters, flower-stalk sometimes partially fused with the stalk of the subtending leaf, bisexual, actinomorphic. KCA perigynous/half epigynous. K5/(5), C5, A5 (G(3); ovules *n*, parietal; styles 3, free, broad, stigmas brush-like. Capsule, seeds with arils. *Mainly tropical America.*

A family of 10 genera and about 100 species. Two genera are native to North America, and 1 species of *Turnera* is occasionally grown as a glasshouse ornamental.

Genera included: ***Piriqueta, Turnera.***

134 (158) Passifloraceae. Trees/shrubs/climbers with tendrils. Leaves alternate, simple/compound, stipulate, stalks often with nectaries; ptyxis conduplicate. Flowers axillary, usually bisexual, actinomorphic, K and C united below, A and G often borne on a stalk (androgynophore). K3–8, usually 5, C usually 5/rarely 0, often with corona, A4–5/(4–5), G(3–5); ovules *n*, parietal;

styles 3–5, united only at the base. Berry/capsule. *Mainly tropical America.*

There are 18 genera and 150 species. One genus is native to North America, and many species of *Passiflora* are grown as ornamental climbers.

Genera included: ***Passiflora.***

135 (159) Cistaceae. Herbs/shrubs. Leaves usually opposite, simple, usually evergreen, stipulate/exstipulate; ptyxis conduplicate or rarely flat. Inflorescence usually cymose, rarely flowers solitary/in racemes. Flowers bisexual, actinomorphic. KCA hypogynous. K3–5, often differing in size and shape, C3–5/rarely more, A*n*, G(3–10) usually (5); ovules 2–*n*, parietal; style 1 or stigmas head-like and sessile. *Mainly warm north temperate areas.*

A family of 7 genera and about 175 species of shrubs or shrublets. Five genera are native to Europe and 4 to North America. Species of a few genera are cultivated as ornamental shrubs.

Genera included: ***Cistus, Fumana, ×Halimiocistus, Halimium, Helianthemum, Hudsonia, Lechea, Tuberaria.***

136 (160) Bixaceae. Trees/shrubs; sap red, orange or golden. Leaves alternate, simple, palmately veined/lobed, stipulate; ptyxis conduplicate. Inflorescences racemes/panicles. KCA hypogynous. K5, imbricate, C5, A*n*, G(2–5); ovules *n*, parietal; style 1. Capsule. *Tropics.*

There is a single genus, native to North America and about 16 species.

Genera included: ***Bixa.***

136a (161) Cochlospermaceae. Trees/shrubs with coloured sap. Leaves alternate, stipulate, palmately lobed. Flowers in panicles or racemes, bisexual, actinomorphic. KCA hypogynous. K5, C5, A*n*, anthers opening by short, pore-like slits, G(3),

1-celled or 3-celled; ovules *n*, parietal; style1. Fruit a capsule, seeds often covered with hairs. *Mainly tropics.*

There are 2 genera, both native to the southern parts of the USA, both rarely cultivated in Europe.

Genera included: ***Amoreuxia, Cochlospermum.***

137 (162) Tamaricaceae. Shrubs/small trees. Leaves usually alternate, usually scale-like and with salt-secreting glands, exstipulate. Flowers in racemes/panicles/rarely solitary, bisexual, actinomorphic. KCA hypogynous. K4–6, C4–6, A4–15, arising from a disc, G(2–5); ovules *n*, parietal/basal; style 1. Capsule, seeds bearded at one end. *Mostly Mediterranean area and Central Asia.*

A family found in rather arid areas, consisting of 5 genera and about 90 species. Two genera are native to Europe and 1 to North America; species of 3 are grown as ornamentals.

Genera included: ***Myricaria, Reaumuria, Tamarix.***

138 (163) Frankeniaceae. Herbs/shrubs. Leaves opposite, entire, exstipulate, often with salt-secreting glands. Inflorescences cymes/flowers solitary. Flowers bisexual, actinomorphic. KCA hypgynous. K(4–7), C4–7, each with an outgrowth (ligule) on its inner face, A4–7/rarely more, usually 6, G(2–4); ovules 2–*n*, parietal; style 1, divided above into 2–4 stigmas. Capsule. *Widespread.*

Another family found in arid, often saline areas; there are 3 genera with about 30 species. One genus is native to both Europe and North America.

Genera included: ***Frankenia.***

139 (164) Elatinaceae. Small herbs, usually aquatic or semi-aquatic. Leaves opposite/whorled, stipulate, simple. Flowers solitary/inflorescences cymose, bisexual, actinomorphic. KCA hypogynous. K2–5, usually 4, C2–5, usually 4, A6–10, usually

8, G(2–5), usually (4); ovules *n*, axile; styles 2–5, usually 4, very small, free. Capsule; seeds pitted. *Widespread.*

Two genera and about 32 species of aquatic herbs. Both genera are native to Europe and North America.

Genera included: ***Bergia, Elatine***.

140 (165) Caricaceae. Soft-wooded trees/shrubs. Leaves alternate, long-stalked, usually divided, exstipulate/stipules spine-like. Flowers solitary/in cymes, usually unisexual, actinomorphic. KCA hypogynous. K(5), C(5), A10 attached to base of C, G(5); ovules *n*, parietal; styles 5, free. Large berry. *Tropical America and Africa.*

There are 4 genera and 31 species. Species of *Carica* are native to North America and a few of them are cultivated in Europe. *Carica papaya* produces the pawpaw of commerce.

Genera included: ***Carica***.

141 (166) Loasaceae. Herbs/shrubs, often with rough/stinging hairs. Leaves alternate/opposite, simple/divided, exstipulate. Flowers solitary and axillary/in cymes, bisexual, actinomorphic. KCA epigynous. K5, C5, A*n*, often united or united in bundles on same radii as petals, G(3–5); ovules *n*, parietal; style 1. Capsule. *Mainly New World.*

There are 15 genera and 250 species; 4 of the genera are native to North America, and a few species are cultivated as ornamentals.

Genera inlcuded: ***Blumenbachia, Caiophora, Cevallia, Eucnide, Loasa, Mentzelia, Petalonyx***.

142 (167) Datiscaceae. Herbs/woody. Leaves alternate, simple/compound, exstipulate. Inflorescence raceme-like. Flowers unisexual, actinomorphic. PA/KCA epigynous. P3–8/K3–8/(3–8), C0–8, A4–*n*/(*n*), G(3), open at apex; ovules *n*, parietal; styles free, sometimes bifid. Capsule. *Scattered.*

A diverse family, with 3 genera and 4 species. One genus is native to Europe and North America. *Datisca cannabina* is occasionally grown as an ornamental.

Genera included: ***Datisca.***

143 (168) Begoniaceae. Herbs/shrubs. Leaves alternate, simple/compound, stipulate, often fleshy, base usually oblique; ptyxis conduplicate or conduplicate-plicate. Flowers usually unisexual, actinomorphic/zygomorphic. PA epigynous, P2–12, usually 4 in 2 unequal pairs in male flowers, A*n*/(*n*), G(2–5), usually (3); ovules *n*, parietal/axile; styles 3, simple or divided. Capsule/berry. *Mostly tropics.*

A uniform family of 3 genera and almost 1400 species. Two genera are native in the southern parts of North America, and species of all 3 (especially *Begonia*) are grown as glasshouse plants, houseplants and bedding annuals.

Genera included: ***Begonia, Hillebrandia, Symbegonia.***

144 (169) Cucurbitaceae. Mostly herbs/climbers with tendrils. Leaves alternate, often lobed, exstipulate; ptyxis conduplicate. Inflorescences axillary cymes/flowers solitary; flowers usually unisexual and actinomorphic. KCA epigynous, K5/(5), C5/(5), A1–5 usually 3/rarely (3), 1 anther 1-celled, G(3–5); ovules *n*, parietal/rarely axile; style 1, stigmas divided, sometimes complex/rarely styles free. Fruit berry-like. *Widespread but mainly tropical.*

A distinctive family of 131 genera and 735 species. Seven genera (4 introduced) occur in Europe and 26 are native to North America. About 28 genera are cultivated; several of them include economically important plants such as *Cucumis* (cucumber, melons) and *Cucurbita* (gourds, marrows, pumpkins).

Genera included: ***Abobra, Apodanthera, Benincasa, Brandegea, Bryonia, Cayaponia, Citrullus, Coccinia, Corallocarpus, Cucumis, Cucurbita, Cyclanthera,***

***Diplocyclos, Ecballium, Echinocystis, Echinopepon, Fevillea, Gurania, Ibervillea, Kedrostis, Lagenaria, Luffa, Marah, Melothria, Momordica, Mukia, Neoalsomitra, Sechium, Sicyos, Sicyosperma, Telfairia, Thladiantha, Trichosanthes, Tumamoca, Xerosicyos, Zehneria*.**

145 (170) Lythraceae. Herbs/shrubs/trees. Leaves opposite/whorled/rarely alternate, stipulate/exstipulate; ptyxis flat or conduplicate. Inflorescences clusters/racemes/panicles. Flowers bisexual, actinomorphic/rarely zygomorphic, often heterostylous. KCA perigynous. K4–6 (epicalyx frequent), C2–6/rarely 0, A6–16/rarely fewer or more, G(2–6); ovules *n*, axile; style 1, stigma capitate. Capsule. *Widespread.*

Twenty-six genera and 580 species. Three genera occur in Europe and 11 are native to North America. Species of 5 or 6 genera are grown as ornamentals.

Genera included: ***Ammannia, Cuphea, Decodon, Ginoria, Heimia, Lagerstroemia, Lawsonia, Lythrum, Nesaea, Rotala*.**

146 (171) Trapaceae. Aquatic annual herb. Leaves opposite below, alternate above, simple, exstipulate, with inflated stalks. Flowers solitary, actinomorphic, bisexual. KCA more or less epigynous. K4, C4, A4, G(2); ovules 1 per cell, axile; style 11. Drupe with spines. *Old World.*

A single genus (native to Europe) with about 30 species; *Trapa natans*, the water chestnut, is cultivated for its edible fruits in some areas.

Genera included: ***Trapa*.**

147 (172) Myrtaceae. Woody. Leaves usually opposite, simple, exstipulate, with translucent aromatic glands; ptyxis flat, conduplicate or supervolute. Inflorescences various. Flowers bisexual, actinomorphic. KCA usually epigynous. K4–5/(4–5), C4–5/(4–5), A*n*/(*n*), G(3–*n*); ovules 2–*n* per cell, axile/parietal; style 1, long, slightly lobed at apex. Capsule/berry. *Mostly tropical America and Australia.*

A large and rather uniform family of 121 genera and over 3800 species. Two genera (1 introduced) occur in Europe and 17 are native to North America. About 32 genera are cultivated for ornament and one of them, *Eucalyptus*, is widely grown in southern Europe.

Genera included: ***Acca, Agonis, Amomyrtus, Baeckea, Beaufortia, Blepharocalyx, Callistemon, Calothamnus, Calyptranthes, Calytrix, Campomanesia, Chamaelaucium, Darwinia, Eremaea, Eucalyptus, Eugenia, Gomidesia, Kunzea, Leptospermum, Lophomyrtus, Lophostemon, Luma, Marlierea, Melaleuca, Metrosideros, Myrceugenia, Myrcia, Myrcianthes, Myrciaria, Myrrhinium, Myrteola, Myrtus, Pimenta, Pseudanamomis, Psidium, Rhodomyrtus, Siphoneugena, Syzygium, Ugni, Verticordia.***

148 (173) Punicaceae. Shrub/small tree, spiny. Leaves opposite or almost so, simple, exstipulate; ptyxis flat. Inflorescence cymose/flowers solitary. Flowers bisexual, actinomorphic. KCA epigynous. K(5–7), C5–7, A*n*, G usually (8–12), cells superposed/rarely (3); ovules *n*, axile; style 1. Fruit berry-like. *Temperate Old World.*

A single genus with 2 species, 1 of them native to Europe. One species (*Punica granatum*) is widely grown as an ornamental and for its edible fruit (pomegranate). Brummitt places the genus in the Lythraceae (above).

Genera included: ***Punica.***

149 (174) Lecythidaceae. Woody. Leaves alternate, simple, usually exstipulate; ptyxis supervolute. Flowers in large spikes, bisexual, actinomorphic/zygomorphic. KCA epigynous. K2–6,

C4–8/rarely 0, A*n*, variously united, G(2–6); ovules 1–*n* per cell, axile; style 1, short, lobed at apex. Fruit leathery or woody, seeds large and woody. *Tropics.*

A family of 20 genera and about 280 species. Species of a few genera are cultivated as ornamentals in greenhouses. *Bertholletia excelsa* produces the Brazil nut.

Genera included: ***Barringtonia, Bertholletia, Couroupita, Lecythis, Napoleonaea.***

150 (175) Melastomataceae. Woody/herbaceous. Leaves opposite/rarely whorled, exstipulate, simple, usually with 3 more or less parallel main veins arising from the base or near it; ptyxis conduplicate or supervolute. Inflorescences usually cymes. Flowers bisexual, perianth actinomorphic. KCA perigynous/epigynous. K3–6, C3–8, A3–*n*/usually 8/10, often unequal, filmaents usually with a conspicuous joint, anthers opening by pores, G(2–5); ovules *n*, axile; style 1. Capsule/berry. *Mainly tropics.*

A large and relatively uniform family of 185 genera and 4000 species. Nineteen genera are native to North America, and species of over 30 are cultivated as ornamentals. Includes Memecylaceae.

Genera included: ***Acisanthera, Amphiblemma, Arthrostemma, Bertolonia, Blakea, Bredia, Calvoa, Calycogonium, Centradenia, Clidemia, Dissotis, Graffenrieda, Gravesia, Henriettella, Heterocentron, Heterotrichum, Maieta, Mecranium, Medinilla, Melastoma, Memecylon, Miconia, Monochaetum, Monolena, Mouriri, Nepsera, Osbeckia, Ossaea, Oxyspora, Pterolepis, Rhexia, Sonerila, Tetrazygia, Tibouchina, Tococa, Triolena.***

151 (176) Rhizophoraceae. Trees/shrubs (mangroves). Leaves alternate/opposite, simple, stipulate (stipules soon falling); ptyxis involute. Inflorescences umbel-like/flowers solitary. Flowers usually bisexual, actinomorphic. KCA epigynous. K usually 4–5,

C usually 4–5, A8–16/*n*, G(2–6); ovules 1 per cell, axile; style 1, short, lobed at tip. Fruit berry-like, seeds often partly developing while still on parent plant. *Tropics.*

There are 16 genera and about 130 species of mangrove. They are rarely seen in gardens except as uncharacteristic juvenile plants.

Genera included: ***Anopyxis, Bruguiera, Cassipourea, Crossostylis, Rhizophora.***

152 (177) Combretaceae. Woody, sometimes climbing or with spines. Leaves alternate/opposite, simple, exstipulate, usually evergreen; ptyxis conduplicate or supervolute. Inflorescences spikes/racemes/panicles. Flowers bisexual/bisexual and male, actinomorphic. KCA epigynous. K4–5, C0/4–5, A8–10, G(2–5), 1-celled; ovules 2–6, apical; style 1, stigma capitate. Fruit 1-seeded, indehiscent, often winged or ridged. *Tropics.*

Nineteen genera and about 500 species. Five genera are native to North America, and species of about 4 are infrequently grown as ornamental shrubs or climbers.

Genera included: ***Anogeissus, Buchenavia, Bucida, Combretum, Conocarpus, Laguncularia, Pteleopsis, Quisqualis, Terminalia.***

153 (178) Onagraceae. Usually herbs. Leaves alternate/opposite, simple, stipulate/exstipulate; ptyxis flat or involute. Flowers solitary/racemose, bisexual, usually actinomorphic. KCA usually epigynous, a tubular epigynous zone often present, K2–6, usually 4, C2–4/rarely 0, A4–8/rarely 1, G(1–5), usually (4); ovules *n*, axile; style 1, lobed in the upper third. Capsule/berry/nut. *Mostly temperate areas.*

There are 24 genera and 650 species. Five genera (1 introduced) occur in Europe and 13 in North America. Species of about 10 genera are grown as ornamentals, the most important being *Fuchsia*, which includes both hardy and glasshouse species.

Genera included: ***Boisduvalia, Calylophus, Camissonia, Chamerion, Circaea, Clarkia, Epilobium, Fuchsia, Gaura, Gayophytum, Heterogaura, Lopezia, Ludwigia, Oenothera, Stenosiphon.***

154 (179) Haloragaceae. Mostly herbs, aquatic or growing in damp places/rarely shrubs/small trees. Leaves alternate/opposite/whorled, stipulate/exstipulate; ptyxis flat or conduplicate-plicate. Flowers solitary/in terminal spikes/panicles, unisexual/bisexual; KCA/PA epigynous. K0–(4), C0–4, A4 or more, usually 8, G(2–4); ovules 1 per cell, axile/apical; styles 1–4, free. Nut/drupe. *Widespread.*

There are 9 genera and about 120 species. One genus occurs in Europe, 5 in North America. A few are grown as ornamentals or aquarium plants.

Genera included: ***Haloragis, Myriophyllum, Proserpinaca.***

154a (180) Gunneraceae. Large, rhubarb-like terrestrial herbs, often growing in marshy places or at water margins. Leaves undivided, stipulate. Flowers unisexual, in panicles. PA (4), A1/2, G1-celled. *Southern hemisphere.*

A single genus with several species cultivated as spectacular water-margin plants.

Genera included: ***Gunnera.***

155 (181) Theligonaceae. Fleshy herb. Lower leaves opposite, upper alternate, all simple and with sheathing bases. Inflorescence cymose. Flowers unisexual; male: actinomorphic, PA hypogynous, P2, A7–22; female: more or less zygomorphic, P tubular, G1, style 1, at last lateral, ovule 1, basal. Nut. *Scattered.*

A single genus (which grows wild in Europe) with 3 species.

Genera included: ***Theligonum.***

156 (182) Hippuridaceae. Aquatic, rhizomatous herb. Leaves whorled, entire, exstipulate. Flowers axillary, unisexual/bisexual, actinomorphic. P0, A1, epigynous; G1; ovule 1, apical; style 1. Cypsela. *Temperate areas.*

A single genus with 4 species, native to both Europe and North America, grown as an aquarium and pond plant.

Genera included: ***Hippuris.***

157 (183) Cynomoriaceae. Root-parasite without chlorophyll. Leaves scale-like. Inflorescence spike- or head-like. Flowers usually unisexual, actinomorphic. PA epigynous. P1–5, A, G1; ovule 1, more or less apical. Small nut. *Mediterranean area to Central Asia.*

A single genus, native to Europe, with 2 species.

Genera included: ***Cynomorium.***

158 (184) Alangiaceae. Trees/shrubs, sometimes spiny, rarely lianas, with latex. Leaves alternate, simple/lobed, exstipulate. Inflorescences axillary cymes. Flowers bisexual, actinomorphic. KCA epigynous. K(4–10) rarely almost 0, C4–10, recurving, A4–20, G(2–3), 1-celled; ovule 1, pendulous; style 1, stigma capitate. Drupe. *Old World tropics and subtropics.*

A single genus with about 18 species, one or two of which are occasionally grown for ornament.

Genera included: ***Alangium.***

159 (185) Nyssaceae. Trees/shrubs. Leaves alternate, simple, exstipulate; ptyxis conduplicate. Flowers in axillary clusters, unisexual/bisexual, actinomorphic. KCA epigynous. K5, C5, A8–12, G(2), 1-celled; ovule 1, axile; style 1, lobed at top. Drupe. *Temperate North America, China.*

There are 2 genera and 7 species; 1 genus is native to North America, and species of both are cultivated. The family is often included in the Cornaceae.

Genera included: ***Camptotheca, Nyssa.***

160 (186) Davidiaceae. Trees. Leaves alternate, simple, exstipulate. Inflorescence a head surrounded by 2 showy white bracts, the head consisting of 1 bisexual flower surrounded by many male flowers. A epigynous in bisexual flower. P0, A1–7, G6–10-celled; ovules 1 per cell, axile; style 1. Drupe. *China.*

A single genus with a single species (the handkerchief tree), widely grown as an ornamental. The family is often included in the *Cornaceae*.

Genera included: ***Davidia***.

161 (187) Cornaceae. Usually woody. Leaves usually opposite/rarely alternate, simple, exstipulate, sometimes evergreen; ptyxis conduplicate or involute. Flowers in corymbs/umbels, unisexual/bisexual, actinomorphic. KCA epigynous, K3–5, C3–5/rarely 0, A3–5, G(2–3); ovules 1 per cell, axile/rarely parietal; style 1, slightly lobed above, or styles 2–3 free. Drupe/berry. *Mostly temperate areas.*

A family of about 10 genera (sometimes reduced to 3 or 4) with over 100 species. *Cornus* is native to both Europe and North America, and many species of it are grown as ornamentals.

Genera included: ***Cornus***.

161a (188) Torricelliaceae. Woody. Leaves alternate, exstipulate. Flowers in panicles, unisexual, actinomorphic. KCA epigynous. Male: K5 small teeth, C5, valvate, A5; female: K3–5-lobed, C0, G(3–4); ovules 1 per cell, apical. Drupe. *Eastern Himalaya, western China.*

A single genus with 3 species, very rarely cultivated.

Genera included: ***Torricellia***.

161b (189) Aucubaceae. Shrubs. Leaves opposite, exstipulate, evergreen, somewhat toothed. Flowers in axillary panicles, unisexual, actinomorphic. KCA epigynous. K4 (minute teeth), C4, brown, A4, G1-celled; ovule 1, apical; style 1. Berry. *Himalaya, eastern Asia.*

A single genus with 3 species, 1 commonly cultivated in shrubberies; its brown petals are unusual.

Genera included: ***Aucuba***.

161c (190) Griseliniaceae. Shrubs. Leaves alternate, exstipulate, evergreen. Flowers in panicles, unisexual, actinomorphic. KCA epigynous. K5 (minute teeth), C5/rarely 0, A5, G(3), 1-celled; ovule 1, apical/ style 1. Berry. *New Zealand, South America*.

One genus with 6 species, one cultivated as a hedging plant and ornamental.

Genera included: ***Griselinia***.

161d (191) Helwingiaceae. Shrubs. Leaves alternate, exstipulate. Flowers in clusters borne apparently on the surfaces of the leaves (by adnation of the inflorescence stalk and the leaf-stalk and midrib), unisexual, actinomorphic. KCA epigynous. K3–5, C3–5, A3–5, G(3–4), 3–4-celled, with 1 ovule in each cell, apical/axile. Drupe. *Himalaya, China, Japan*.

A single genus with 3 species, one occasionally cultivated as a curiosity.

Genera included: ***Helwingia***.

162 (192) Garryaceae. Evergreen shrubs/trees. Leaves opposite, entire, exstipulate; ptyxis conduplicate. Inflorescences catkin-like. Flowers unisexual; male: P4, A4; female: P0, G(2), naked/inferior, 1-celled; ovules 2, apical; styles 2, free, stigmas lateral. Berry. *North America*.

A single genus with about 8 species, native to North America. Several species are grown as ornamental trees and shrubs. The inflorescences qualify technically as catkins, but this family is very different from those others with catkinate inflorescences (e.g. Betulaceae, Fagaceae).

Genera included: ***Garrya***.

163 (193) Araliaceae. Herbs/shrubs/trees, often spiny. Leaves alternate, usually lobed/compound, stipulate; stellate hairs frequent; ptyxis mainly conduplicate. Inflorescences umbels, these often aggregated into complex panicles. Flowers unisexual/bisexual, actinomorphic. KCA epigynous. K4–5/(4–5)/rim-like, C4–15, usually 5, usually valvate, A4–15, usually 5, G(2–30); ovules 1 per cell, axile; style single or styles free and as many as ovary-cells. Berry/drupe. *Mainly tropics.*

A fairly uniform family of about 25 genera and over 1400 species. One genus is native to Europe and 11 to North America; species of most genera are in cultivation.

Genera included: ***Aralia, Cheirodendron, Cussonia, Dendropanax, Didymopanax, Eleutherococcus, ×Fatshedera, Fatsia, Hedera, Kalopanax, Meryta, Metapanax, Munroidendron, Oplopanax, Oreopanax, Panax, Polyscias, Pseudopanax, Reynoldsia, Schefflera, Tetrapanax, Tetraplasandra, Trevesia.***

164 (194) Umbelliferae/Apiaceae. Usually aromatic herbs. Leaves alternate, often pinnately compound, stalks sheathing the shoots but without stipules; ptyxis conduplicate or supervolute. Inflorescences umbels/rarely heads. Flowers usually bisexual, actinomorphic, sometimes some of the outer flowers of the umbel zygomorphic. KCA epigynous. K(5), often very reduced, C5, imbricate and inflexed, A5, G(2); ovules 1 per cell, axile; styles 2, borne on a swollen stylopodium. Schizocarp. *Widespread.*

A large and uniform family with 420 genera and 3100 species. One hundred and ten genera are native to Europe and 83 to North America. Species of many genera are cultivated for ornament, some as vegetables, e.g. carrots (*Daucus carota*), celery (*Apium graveolens* var. *dulce*), and parsnip (*Pastinaca sativa*), or as herbs and flavourings, e.g. caraway (*Carum carvi*), coriander (*Coriandrum sativum*), cumin (*Cuminum cyminum*), etc.

Genera included: ***Aciphylla, Actinotus, Aegopodium, Aethusa, Ainsworthia, Aletes, Ammi, Ammoides, Ammoselinum, Anethum, Angelica, Anisotome, Anthriscus, Apiastrum, Apium, Artedia, Astrantia, Astrodaucus, Athamanta, Aulacospermum, Azorella, Berula, Bifora, Bolax, Bonannia, Bowlesia, Bunium, Bupleurum, Cachrys, Capnophyllum, Carum, Caucalis, Cenolophium, Centella, Chaerophyllum, Cicuta, Cnidium, Conioselinum, Conium, Conopodium, Coriandrum, Crithmum, Cryptotaenia, Cuminum, Daucus, Dethawia, Echinophora, Elaeoselinum, Endressia, Erigenia, Eriosynaphe, Eryngium, Eurytaenia, Falcaria, Ferula, Ferulago, Foeniculum, Glehnia, Glia, Grafia, Guillonea, Hacquetia, Harbouria, Heptapera, Heracleum, Hladnikia, Hohenackeria, Huetia, Hydrocotyle, Johrenia, Kundmannia, Lagoecia, Laser, Laserpitium, Lecokia, Levisticum, Ligusticum, Lilaeopsis, Limnosciadium, Lomatium, Magydaris, Malabaila, Melanoselinum, Meum, Molopospermum, Musineon, Myrrhis, Myrrhoides, Naufraga, Neoparrya, Oenanthe, Opopanax, Oreonana, Oreoxis, Orlaya, Orogenia, Osmorhiza, Oxypolis, Palimbia, Pastinaca, Perideria, Petroselinum, Peucedanum, Phlojodicarpus, Physospermum, Pimpinella, Pleurospermum, Podistera, Polytaenia, Portenschlagiella, Pseudorlaya, Pseudotaenidia, Ptilimnium, Ptychotis, Ridolfia, Rouya, Sanicula, Scaligeria, Scandix, Sclerochorton, Selinum, Seseli, Silaum, Sison, Sium, Smyrnium, Spermolepis, Sphenosciadium, Stefanoffia, Taenidia, Tauschia, Thapsia, Thaspium, Thorella, Tilingia, Tordylium, Torilis, Trachymene, Trepocarpus, Trinia, Trochiscanthes, Turgenia, Turgeniopsis, Xatardia, Yabea, Zizia, Zosima.***

165 (195) Diapensiaceae. Evergreen herbs/shrublets. Leaves alternate, simple, exstipulate; ptyxis conduplicate. Flowers solitary or in heads/racemes, bisexual, actinomorphic. K hypogynous, CA perigynous. K(5), C(5), A5 with 5–0 staminodes, G(3);

ovules usually *n*, axile; style 1, short, 3-lobed above. Capsule. *North temperate areas.*

A small family of 6 genera and 13 species. One genus is native to Europe, 4 to North America. Three of the genera are cultivated as ornamentals.

Genera included: ***Diapensia, Galax, Pyxidanthera, Shortia.***

166 (196) Clethraceae. Shrubs/trees, evergreen /deciduous. Leaves alternate, simple, exstipulate, with stellate hairs; ptyxis supervolute. Inflorescences racemes/panicles. Flowers fragrant, bisexual, actinomorphic, disc 0. KCA hypogynous. K(5–6), C5–6. A10–12, sometimes slightly attached to bases of petals, G(3); ovules *n* per cell, axile; style 1, stigma 3-lobed. Capsule. *Mostly tropics and subtropics.*

There is a single genus with about 64 species, some of which are native to North America. About 7 species are cultivated as ornamental shrubs.

Genera included: ***Clethra.***

167 (197) Pyrolaceae. Herbs/shrublets, sometimes saprophytic. Leaves alternate, often in rosettes, evergreen, exstipulate. Flowers solitary/in racemes, bisexual, actinomorphic. KCA hypogynous. K4–5, C4–5, A8–10, anthers opening by pores, G(4–5); ovules *n* per cell, axile/parietal; style 1, stigmas divided at apex. Capsule/berry. *North temperate areas.*

Often included in Ericaceae, there are 4 genera, all found in Europe, and about 42 species. Several species from 3 of the genera are cultivated as ornamentals, even though they are difficult to grow well.

Genera included: ***Chimaphila, Moneses, Orthilia, Pyrola.***

168 (198) Ericaceae. Woody/herbs. Leaves alternate/opposite/ appearing whorled/basal, simple, exstipulate, usually evergreen, sometimes needle-like; ptyxis very variable. Inflorescences racemes/clusters/flowers solitary. Flowers bisexual, actinomorphic or rarely zygomorphic. KCA hypogynous/epigynous, rarely K hypogynous, CA perigynous, disc present. K(4–5)/rarely 4–5, sometimes very small, C(3–5)/3–5/rarely –10, A5–10/rarely –27, anthers usually opening by pores, pollen in tetrads, G(2–12); ovules *n*, axile/rarely parietal; style 1, lobed at apex/stigmas enclosed in a sheath. Capsule/berry/drupe. *Widespread.*

A large and diverse family with about 100 genera and over 3000 species. Eighteen genera occur in Europe and over 40 in North America. Species of many of the genera are cultivated as ornamentals, especially the 2 largest, *Rhododendron* and *Erica*. Those genera with inferior ovaries were formerly separated off as the Vacciniaceae.

Genera included: ***Agapetes, Agarista, Agauria, Andromeda, Arbutus, Arctostaphylos, Bejaria, Bruckenthalia, Bryanthus, Calluna, Cassiope, Cavendishia, Chamaedaphne, Cladothamnus, Comarostaphylis, Craibiodendron, Daboecia, Elliottia, Enkianthus, Epigaea, Erica, Gaultheria, Gaylussacia, Gonocalyx, Kalmia, Kalmiopsis, Ledum, Leiophyllum, Leucothoe, Loiseleuria, Lyonia, Macleania, Menziesia, Ornithostaphylis, Oxydendrum, Pernettya, ×Phylliopsis, Phyllodoce, ×Phyllothamnus, Pieris, Pleuricospora, Pterospora, Rhododendron, Rhodothamnus, Sarcodes, Symphysia, Tripetaleia, Tsusiophyllum, Vaccinium, Xylococcus, Zenobia.***

168a (199) Monotropaceae. Parasitic or saprophytic herbs without chlorophyll. Leaves reduced to scales, alternate, exstipulate. Flowers in heads /solitary, bisexual, actinomorphic. KCA hypogynous. K2–6, very bract-like, C3–6/(3–6), A6–12, anthers opening by longitudinal slits, pollen mostly in a mass, G1–6-celled, ovules numerous, axile/parietal; style 1, stigmas enclosed within a sheath. Capsule with copious minute seeds. *Northern hemisphere.*

There are 13 genera and fewer than 20 species. Two genera are native to Europe, 4 to North America.

Genera included: ***Allotropa, Monotropa, Monotropsis, Pityopus.***

169 (200) Empetraceae. Evergreen, heath-like shrublets. Leaves alternate/almost whorled, entire, exstipulate. Flowers solitary/in terminal clusters, unisexual/bisexual, actinomorphic. PA hypogynous, disc 0. P2–6, A3–4, G(2–9) ovules 1 per cell, axile; style 1 with 2–9 branches above. Drupe with 2–9 stones. *Temperate areas.*

Three genera and 5 species with a remarkable distribution in both north and south temperate areas. Two genera are native to Europe, 3 to North America.

Genera included: ***Ceratiola, Corema, Empetrum.***

170 (201) Epacridaceae. Shrubs/small trees. Leaves alternate, rigid, simple, exstipulate, evergreen; ptyxis flat. Inflorescences spikes/racemes/panicles. Flowers bisexual, actinomorphic. K hypogynous, CA perigynous. K(4–5), C(4–5), A4–5, anthers 1-celled, G(2–10); ovules 1–*n* per cell, axile; style 1. Capsule/drupe. *Mainly Australasia.*

A rather uniform family of 31 genera and 400 species, particularly well developed in Australia. Species of a few of the genera are cultivated.

Genera included: ***Cyathodes, Epacris, Richea, Styphelia, Trochocarpa.***

171 (202) Theophrastaceae. Shrubs. Leaves alternate, sometimes in false terminal whorls, simple, exstipulate; ptyixs conduplicate. Flowers solitary/paired/in racemes, usually bisexual, actinomorphic. K hypogynous, CA perigynous. K(4–6) usually 5, C(4–6) usually 5, A4–6 usually 5, on the same radii as the C-lobes, anthers opening by pores towards the outside of the flower, with usually 5 staminodes, G1-celled; ovules *n*, free-central; style simple, lobed at apex. Drupe. *New World tropics*.

There are 6 genera and about 100 species; 1 genus occurs in North America, and species of all 4 genera are grown as glasshouse ornamentals.

Genera included: ***Clavija, Deherainia, Jacquinia, Theophrasta***.

172 (203) Myrsinaceae. Woody. Leaves alternate/rarely whorled, exstipulate, mostly evergreen, with translucent or coloured dots or stripes; ptyxis variable, mainly supervolute. Inflorescences cymose/clusters. Flowers unisexual/bisexual, usually actinomorphic. K hypogynous, CA perigynous/rarely KCA half-epigynous. K(4–5), C(4–5)/rarely 4–5, A4–5 on same radii as C-lobes, opening by slits towards the inside of the flower, G(4–5); ovules *n*, free-central; style 1, stigmas 4–5. Berry/drupe. *Mainly tropics*.

A fairly uniform family of 39 genera and about 1250 species. One genus is native to Europe, 8 to North America. Species of about 4 genera are occasionally grown as ornamental shrubs. Includes Maesaceae.

Genera included: ***Ardisia, Embelia, Maesa, Myrsine, Parathesis, Rapanea, Stylogyne, Wallenia***.

173 (204) Primulaceae. Herbs/rarely shrublets, rarely aquatic. Leaves alternate/opposite/basal, usually simple, exstipulate; ptyxis variable. Inflorescence various, often superposed whorls. Flowers bisexual, actinomorphic/rarely zygomorphic. K hypogynous, CA perigynous/rarely KCA half-epigynous/very rarely PA hypogynous. K(5–7), C(5–7)/rarely 0, A5–7, on same radii as C-lobes, G usually (5), 1-celled; ovules *n*, free-central; style 1, stigma capitate. Capsule. *Widespread*.

A uniform family of 22 genera and about 800 species. Fourteen genera are native to Europe, 10 to North America. Many of the species are cultivated, especially those belonging to the largest genus, *Primula*. Includes Coridaceae and Samolaceae.

Genera included: ***Anagallis, Androsace, Ardisiandra, Coris, Cortusa, Cyclamen, Dionysia, Dodecatheon, Douglasia, Glaux, Hottonia, Lysimachia, Omphalogramma, Primula, Samolus, Soldanella, Trientalis, Vitaliana.***

174 (205) Plumbaginaceae. Herbs/shrubs/climbers. Leaves alternate/basal, simple, exstipulate; ptyxis flat or involute. Inflorescence racemose/cymose/ flowers often aggregated in 'spikelets'. Flowers bisexual, actinomorphic. KCA hypogynous/K hypogynous, CA perigynous. K5/(5), C(5)/rarely 5, A5 on same radii as C-lobes, G(5), 1-celled, ovule 1, basal; styles 5, free. Fruit indehiscent, retained in K-tube. *Widespread.*

Another relatively uniform family in which between 15 and 27 genera are recognised by different authors; there are over 500 species, many of them growing only in brackish areas. Eight genera are native to Europe, 3 to North America. Species of about 5 are cultivated as ornamental herbaceous plants; several have 'everlasting' flowers and are grown for the cut-flower trade.

Genera included: ***Acantholimon, Armeria, Ceratostigma, Goniolimon, Limoniastrum, Limonium, Plumbago, Psylliostachys.***

175 (206) Sapotaceae. Trees/shrubs/woody climbers, with milky sap, sometimes spiny. Leaves alternate/rarely opposite/whorled, simple, usually exstipulate, leathery; ptyxis conduplicate. Flowers solitary/in clusters, actinomorphic, unisexual/ bisexual. K hypogynous, CA perigynous. K4–11/(4–6), C(4–18), A4–43, often some staminodial, G(1–30); ovules usually 1 per cell, axile; style 1, stigma capitate/lobed. Berry/drupe. *Mostly tropics.*

There are between 53 and 100 genera (depending on interpretation) and 1100 species. Seventeen genera are native to North America and species of about 11 are occasionally cultivated.

Genera included: ***Chrysophyllum, Diploknema, Madhuca, Manilkara, Micropholis, Mimusops, Nesoluma, Palaquium, Payena, Pouteria, Sideroxylon, Synsepalum, Vitellariopsis.***

176 (207) Ebenaceae. Woody, wood dark, sap watery. Leaves alternate, simple, entire, leathery, exstipulate; ptyxis conduplicate or supervolute. Flowers solitary/in cymes. K hypogynous, CA perigynous, actinomorphic, usually unisexual. K(4), C(4), A6–20, G(2–16); ovules 1–2 per cell, axile, pendulous; style 1, stigma capitate or style short, divided above into as many branches as there are ovary-cells. Berry. *Mostly tropics.*

There are between 2 and 5 genera and about 500 species. One genus is introduced in Europe and 1 is native in North America. A few species of *Diospyros* are cultivated as ornamental trees; ebony is the wood from *D. ebenum* and perhaps other species.

Genera included: ***Diospyros, Euclea.***

177 (208) Styracaceae. Trees/shrubs. Leaves alternate, simple, exstipulate, with stellate hairs or scales; ptyxis mainly supervolute. Inflorescences panicles/racemes/clusters. Flowers bisexual, actinomorphic. KCA hypogynous/epigynous/K hypogynous CA perigynous. K(2–10), C(4–8), A4–8/8–16, G(2–5); ovules *n*, axile; style 1, stigma capitate or 2–5-lobed. Drupe/capsule. *East Asia, America, Mediterranean area.*

A family of 12 genera and 165 species of scattered distribution. One genus is native to Europe and 2 to North America. Species of 5 of the genera are cultivated as small ornamental trees.

Genera included: ***Halesia, Pterostyrax, Rehderodendron, Sinojackia, Styrax.***

178 (209) Symplocaceae. Trees/shrubs. Leaves alternate, simple, exstipulate, leathery. Flowers in spikes/racemes/panicles/ rarely solitary, bisexual, actinomorphic. KCA more or less epigynous. K(4–5), C(4–10), A4–n, G(2–5): ovules 2–4 per cell, axile; style 1. Berry/drupe. *Tropical America, Asia, Australasia.*

A single genus with about 250 species, a few of which are grown as ornamentals.

Genera included: ***Symplocos.***

179 (210) Oleaceae. Woody, sometimes climbing. Leaves usually opposite, simple/pinnately compound, exstipulate; ptyxis flat, conduplicate or supervolute. Inflorescence of cymose panicles. Flowers usually bisexual, actinomorphic. KCA hypogynous/K hypogynous CA perigynous. K(4)/rarely 0(–15), C(4)/rarely 0(–15), A2, G(2); ovules usually 2 per cell, axile; style 1 divided above into 2 stigmas. Fruit various. *Temperate areas and tropics.*

An easily recognised family of 24 genera and over 500 species. Nine genera are native to Europe and 12 to North America. Many species of several genera are grown as ornamental shrubs or trees. *Olea europaea* produces the olive.

Genera included: ***Abeliophyllum, Chionanthus, Fontanesia, Forestiera, Forsythia, Fraxinus, Haenianthus, Jasminum, Ligustrum, Menodora, Olea, Osmanthus, Phillyrea, Picconia, Syringa.***

180 (211) Loganiaceae. Woody, sometimes climbing, with internal phloem; glandular hairs absent. Leaves opposite, entire, stipulate, ptyxis flat. Inflorescence cymose/flowers solitary. Flowers bisexual, actinomorphic. K hypogynous, CA perigynous. K(4–5), imbricate, C(4–5), A4–5, G(2): ovules *n*, axile; style 1, stigma capitate or bilobed. Capsule/berry/drupe. *Tropics.*

A mainly tropical family of 29 genera and 800 species. Five genera are native in North America, and species of a few more are cultivated. The Buddlejaceae (see below, p. 218) is sometimes included in this family. Includes *Gelsemiaceae.*

Genera included: ***Gelsemium, Spigelia.***

181 (212) Desfontainiaceae. Shrubs. Leaves opposite, exstipulate, evergreen, spiny-margined; ptyxis conduplicate. Flowers solitary/inflorescences cymose, bisexual, actinomorphic. K hypogynous, CA perigynous. K5, spine-like, C(5) with long tube, A5, G(5); ovules *n*, axile; style 1. Berry. *Andes.*

A single genus and species cultivated in gardens for its holly-like foliage and large, tubular, red flowers. Often included in the Loganiaceae.

Genera included: ***Desfontainia***.

182 (213) Gentianaceae. Herbs, sometimes climbing/small shrubs. Leaves mostly opposite/rarely whorled/alternate, simple, exstipulate; ptyxis variable, often supervolute. Flowers solitary or in cymes/panicles, bisexual, actinomorphic/rarely zygomorphic. K hypogynous, CA perigynous. K(4–5)/rarely (–12), C(4–5)/rarely (–12), A4–5, rarely –12, anthers opening by slits/rarely pores, G(2); ovules *n*, usually parietal; style 1 divided above/stigmas 2, sessile. Capsule/berry-like. *Widespread.*

A uniform family of 80 genera and over 700 species. Nine genera are native in Europe, 17 in North America. Species of about 12 genera are cultivated.

Genera included: ***Bartonia, Blackstonia, Centaurium, Chironia, Cicendia, Crawfurdia, Enicostema, Eustoma, Exaculum, Exacum, Frasera, Gentiana, Gentianella, Gentianodes, Gentianopsis, Halenia, Lisianthius, Lomatogonium, Obolaria, Orphium, Sabatia, Schultesia, Swertia, Tripterospermum***.

183 (214) Menyanthaceae. Aquatic/marsh herbs. Leaves alternate, entire/trifoliolate, stalks sheathing; ptyxis involute or supervolute. Inflorescence various. Flowers bisexual, actinomorphic. K hypogynous, CA perigynous. K(5)/5, C(5), valvate, A5, G(2); ovules *n*, parietal; style 1, bifid into 2 stigmas at apex. Capsule. *Temperate areas.*

Five genera with 40 species, sometimes included in the Gentianaceae. Two genera are native to Europe and 2 to North America; species of 2 are cultivated as ornamental aquatics.

Genera included: ***Fauria, Menyanthes, Nephrophyllidium, Nymphoides.***

184 (215) Apocynaceae. Woody/herbs, often climbing, with milky sap. Leaves entire, usually opposite, exstipulate; ptyxis flat, conduplicate or involute. Inflorescences racemose/cymose/flowers solitary. Flowers bisexual, actinomorphic. K hypogynous, CA perigynous. K4–5/(4–5), C(5) contorted, A5, G(2), often the bodies of the carpels free, united only by the single style; ovules *n*, marginal/axile. Fruits various, generally follicles, seeds often plumed. *Widespread.*

A rather uniform family of 215 genera and 2100 species. Four genera are native to Europe, 27 to North America; species of about 12 genera are cultivated as ornamentals. In some recent accounts, this family is merged with the next.

Genera included: ***Acokanthera, Adenium, Allamanda, Alstonia, Alyxia, Amsonia, Anechites, Angadenia, Apocynum, Beaumontia, Carissa, Catharanthus, Cerbera, Echites, Forsteronia, Funtumia, Haplophyton, Holarrhena, Mandevilla, Nerium, Ochrosia, Pachypodium, Plumeria, Prestonia, Pteralyxia, Rauvolfia, Rhabdadenia, Rhazya, Strophanthus, Tabernaemontana, Thevetia, Trachelospermum, Vallesia, Vinca.***

185 (216) Asclepiadaceae. Woody/herbs/climbers, some succulent, usually with milky sap. Leaves opposite, entire, stipules minute/0; ptyxis flat. Inflorescences racemose/cymose/flowers solitary. Flowers bisexual, actinomorphic. K hypogynous CA perigynous. K(5)/5, C(5) contorted, corona frequent, A5, often joined to styles; 'translators' and pollinia frequent, G(2), carpels often free, united by the common style; ovules *n*, marginal/axile;

style 1, large, head-like. Fruit of 1–2 follicles, seeds plumed. *Mostly tropics.*

Very similar to the Apocynaceae, this family includes 348 genera and 2900 species. Eight genera are native to Europe, 15 to North America. Species of a wide range of genera are grown as ornamentals, some of them as succulents.

Genera included: ***Araujia, Asclepias, Brachystelma, Calotropis, Caralluma, Ceropegia, Cionura, Cynanchum, Dischidia, Dregea, Duvalia, Echidnopsis, Edithcolia, Fockea, Frerea, Gomphocarpus, Gonolobus, Hoodia, ×Hoodiella, Hoya, Huernia, Huerniopsis, Lavrania, Marsdenia, Matelea, Metaplexis, Morrenia, Orbea, Orbeanthus, Oxypetalum, Pachycymbium, Periploca, Piaranthus, Quaqua, Sarcostemma, Stapelia, Stapelianthus, Stephanotis, Tridentea, Tromotriche, Vincetoxicum.***

186 (217) Rubiaceae. Herbs/shrubs/rarely scramblers. Leaves opposite/whorled, stipulate, stipules sometimes leaf-like; ptyxis mainly flat, occasionally revolute or supervolute. Inflorescences various. Flowers usually bisexual, usually actinomorphic. KCA epigynous/half-epigynous. K(4–5)/rarely (–12), C(4–5)/rarely (–12), A4–5, rarely (–12), G(2–5), ovaries of several flowers sometimes coalescing; ovules 1–*n* per cell, axile; style 1, divided above/2, free. Capsule/berry/schizocarp. *Widespread.*

A very large family, with 631 genera and 14 000 species. Nine genera are native to Europe and 60 to North America. Species of about 34 genera are cultivated.

Genera included: ***Alberta, Antirhea, Asperula, Bobea, Bouvardia, Callipeltis, Canthium, Catesbaea, Cephalanthus, Chiococca, Chione, Cinchona, Coccocypselum, Coffea, Coprosma, Crucianella, Cruciata, Diodia, Emmenopterys, Erithalis, Ernodea, Exostema, Faramea, Galium, Gardenia, Genipa, Geophila, Gonzalagunia, Gouldia, Guettarda, Hamelia, Hedyotis,***

Hillia, Hoffmannia, Houstonia, Hydrophylax, Ixora, Kelloggia, Lasianthus, Leptodermis, Luculia, Lucya, Machaonia, Manettia, Mitchella, Mitracarpus, Mitriostigma, Morinda, Mussaenda, Nertera, Paederia, Palicourea, Pentas, Pentodon, Phuopsis, Pinckneya, Psychotria, Putoria, Randia, Rondeletia, Rothmannia, Rubia, Sabicea, Schradera, Sclolosanthus, Serissa, Sherardia, Spermacoce, Strumpfia, Valantia.

187 (218) Polemoniaceae. Herbs/rarely shrubs. Leaves alternate/opposite, entire/pinnately divided, usually exstipulate; ptyxis conduplicate. Inflorescence cymose to head-like/rarely flowers solitary. Flowers bisexual, usually actinomorphic. K hypogynous, CA perigynous. K(5), C(5) contorted, A5, G(3); ovules 1–*n* per cell, axile; style 1, divided above into 3 stigmas. Capsule. *Mainly America.*

There are 20 genera with 275 species. Two genera (1 introduced) are found in Europe and 14 are native to North America. Species of about 8 are grown as ornamental herbaceous plants.

Genera included: ***Allophyllum, Cantua, Collomia, Eriastrum, Gilia, Gymnosteris, Ipomopsis, Langloisia, Leptodactylon, Linanthus, Loeselia, Navarettia, Phlox, Polemonium.***

187a (219) Cobaeaceae. Woody climbers. Leaves alternate, pinnate, the terminal pair of leaflets replaced by hooked tendrils, exstipulate but the lowermost leaflets appearing like stipules. Flowers solitary or a few together on a long stalk; flower-stalks twisting after the flower opens. Flowers bisexual, actinomorphic, disc present. K hypogynous, CA perigynous. K5, C(5), contorted, A5, G(3); ovules 2 or more in each cell, axile; style 1 divided into 3 above. Capsule. *Central and South America.*

There is a single genus with 10–19 species, one of which is often grown as a vigorous climber.

Genera included: ***Cobaea.***

188 (220) Fouquieriaceae. Woody, spiny. Leaves alternate, simple, exstipulate, fleshy. Inflorescence a terminal panicle. Flowers bisexual, actinomorphic. K hypogynous, CA perigynous. K5, C(5), A10–17, G(3); ovules 12–18, parietal; style 1, stigma 3-lobed at the apex. Capsule. *Central and southwest North America.*

A single genus with 11 species of succulent plants, most of them found in western North America, rarely cultivated in Europe.

Genera included: ***Fouquieria.***

189 (221) Convolvulaceae. Climbers/shrublets, often with milky sap/rarely twining parasites without chlorophyll. Leaves alternate, simple, exstipulate, scale-like in parasites; ptyxis conduplicate. Inflorescence cymose/clustered/flowers solitary. Flowers bisexual, mostly actinomorphic. K hypogynous, CA perigynous. K4–5/(4–5), C(4–5), contorted and often folded down the central line of each lobe, A5, sometimes with scales below their insertion, G(2), sometimes 4-celled; ovules 1–2 per cell, axile/parietal; styles 2, free or united and stigma capitate. Capsule/fleshy. *Widespread.*

A relatively uniform family of 58 genera and 1650 species. Six genera are native to Europe, 18 to North America. Species of about 8 genera are cultivated for ornament. The parasitic genus *Cuscuta* is sometimes separated off as the family Cuscutaceae.

Genera included: ***Aniseia, Argyreia, Bonamia, Calystegia, Convolvulus, Cressa, Cuscuta, Dichondra, Evolvulus, Hildebrandtia, Ipomoea, Jacquemontia, Merremia, Operculina, Petrogenia, Porana, Stictocardia, Stylisma, Turbina.***

190 (222) Hydrophyllaceae. Herbs/rarely shrubs, occasionally with irritant or stinging hairs. Leaves alternate/basal/rarely opposite, entire/divided, exstipulate; ptyxis flat or conduplicate. Inflorescences coiled cymes/flowers solitary. Flowers bisexual, actinomorphic. K hypogynous, CA perigynous. K5/(5), C(5) usually imbricate, A5, G(2); ovules *n*/rarely 4, parietal/rarely axile; style 1 or 2, free or variously united below. Capsule. *Mainly America.*

A smallish family with 22 genera and 275 species. Seventeen genera are native to North America, and 2 of them are introduced in Europe. Species of 9 genera are cultivated for ornament, members of the genus *Phacelia* being most often seen. Includes Hydroleaceae.

Genera included: ***Draperia, Ellisia, Emmenanthe, Eriodictyon, Eucrypta, Hesperochiron, Hydrolea, Hydrophyllum, Lemmonia, Nama, Nemophila, Phacelia, Pholistoma, Romanzoffia, Tricardia, Turricula, Wigandia.***

191 (223) Boraginaceae. Herbs/woody, stems circular in section. Leaves alternate, simple, exstipulate; ptyxis conduplicate or supervolute. Inflorescences often coiled cymes. Flowers bisexual, actinomorphic/rarely zygomorphic. K hypogynous, CA perigynous. K5/(5), C(5), A5, G(2) usually 4-celled by the growth of secondary septa; ovules 4, side-by-side, axile; style 1, terminal or more commonly arising between the 4 ovary-cells; stigma simple or slightly 2-lobed. Fruit 4–1 nutlets/drupe. *Widespread.*

A large family with 156 genera and 2500 species. Thirty-four genera are native to Europe, 34 to North America. Species of many are cultivated as ornamentals. *Cordia* and *Ehretia* are sometimes separated off as distinct families (Cordiaceae, Ehretiaceae).

Genera included: ***Alkanna, Amsinckia, Anchusa, Antiphytum, Argusia, Arnebia, Asperugo, Borago, Bothriospermum, Bourreria, Brunnera, Buglossoides, Caccinia, Cerinthe, Cordia, Cynoglossum, Cynoglottis, Echium, Ehretia, Elizaldia, Eritrichium, Hackelia, Halacsya, Harpagonella, Heliotropium, Lappula, Lindelofia, Lithodora, Lithospermum, Lobostemon, Macromeria, Mertensia, Moltkia, Myosotidium, Myosotis, Neatostema, Nonea, Omphalodes, Onosma, Onosmodium, Pectocarya, Pentaglottis, Plagiobothrys, Pulmonaria, Rindera, Rochefortia, Rochelia, Solenanthus, Symphytum, Tiquilia, Tournefortia, Trachystemon, Trigonitis.***

192 (224) Lennoaceae. Parasitic herbs, chlorophyll 0. Leaves scale-like. Inflorescence spicate/corymbose/head-like. Flowers bisexual, actinomorphic. K hypogynous, CA perigynous. K(6–10), C(5–8), imbricate, A5–8, G(6–15); ovules 2 per cell, axile; style 1. Fleshy capsule. *SW USA, Mexico.*

Three genera and 6 species, 2 genera native to North America.

Genera included: ***Ammobroma, Pholisma.***

193 (225) Verbenaceae. Woody/herbaceous. Leaves opposite, simple/compound, exstipulate; ptyxis mainly conduplicate. Inflorescences various. Flowers bisexual, zygomorphic. K hypogynous, CA perigynous. K(5–8) more or less actinomorphic, C(5), A4/rarely 2–5, G2–9-celled, style 1, terminal, divided into 2–9 stigmas at apex; ovules 1–2 per cell, axile/rarely parietal. Drupe/berry/rarely 4 nutlets. *Mainly tropics.*

A traditional interpretation of this family is used here; there are 91 genera and 1900 species. Four genera (1 introduced) occur in Europe and 23 in North America. Species of about 25 genera are cultivated for ornament. The distinction between this family and Labiatae is not clear and genera may be assigned differently.

Genera included: ***Aegiphila, Aloysia, Amasonia, Bouchea, Callicarpa, Caryopteris, Castelia, Citharexylum, Clerodendrum, Diostea, Duranta, Glandularia, Gmelina, Holmskioldia, Lantana, Lippia, Petitia, Petrea, Phyla, Priva, Rhaphithamnus, Stachytarpheta, Stylodon, Verbena, Vitex.***

194 (226) Callitrichaceae. Aquatic herbs. Leaves opposite, simple, exstipulate. Flowers solitary, axillary, unisexual, actinomorphic. A hypogynous. P0, A1, G(2) 4-celled by secondary septa; ovules 1 per cell, axile; styles 2, free. Schizocarp. *Widespread.*

The single genus contains 17 or more species and is native to both Europe and North America. A few species are cultivated as aquarium plants.

Genera included: ***Callitriche.***

195 (227) Labiatae/Lamiaceae. Herbs/shrubs, stems generally square in section. Leaves opposite, aromatic, simple/compound, exstipulate; ptyxis variable but mainly conduplicate. Flowers usually in verticils, mostly bisexual, zygomorphic. K hypogynous, CA perigynous. K usually (5), often zygomorphic, C(5)/rarely (3), 1–2-lipped, A4/2, G(2), 4-celled by secondary septa; style 1, usually arising between the 4 units of the ovary/rarely terminal, divided into 2 stigmas above; ovules 1 per cell, axile, side-by-side. Fruit of 4 nutlets/rarely fleshy. *Widespread.*

A rather uniform family of 224 genera and 5600 species. Forty-one genera are found in Europe and 70 in North America. Species of about 70 genera are cultivated as ornamentals or as aromatic herbs, e.g. mint (*Mentha*), rosemary (*Rosmarinus*), sage (*Salvia*) and thyme (*Thymus*).

Genera included: ***Acanthomintha, Acinos, Agastache, Ajuga, Ballota, Blephilia, Brazoria, Calamintha, Cedronella, Chelonopsis, Cleonia, Clinopodium, Collinsonia, Colquhounia, Conradina, Cunila, Dicerandra, Dracocephalum, Elsholtzia, Eremostachys, Galeopsis, Glechoma, Haplostachys, Hedeoma, Horminum, Hyptis, Hyssopus, Lallemantia, Lamium, Lavandula, Leonotis, Leonurus, Lepechinia, Leucas, Lycopus, Macbridea, Marrubium, Marsypianthes, Meehania, Melissa, Melittis, Mentha, Micromeria, Molucella, Monarda, Monardella, Mosla, Nepeta, Ocimum, Origanum, Perilla, Perovskia, Phlomis, Phyllostegia, Physostegia, Pilophlebis, Plectranthus, Pogogyne, Pogostemon, Poliomintha, Prasium, Prostanthera, Prunella, Pycnanthemum, Pycnostachys, Rhododon, Rosmarinus, Rostrinucula, Salvia, Satureja, Scutellaria, Sideritis, Solenostemon, Stachydeoma, Stachys, Stenogyne, Synandra, Tetraclea, Tetradenia, Teucrium, Thymbra, Thymus, Trichostema, Westringia, Wiedemannia, Ziziphora***.

196 (228) Nolanaceae. Herbs/shrublets. Leaves alternate, simple, exstipulate, fleshy. Flowers axillary, bisexual, actinomorphic.

K hypogynous, CA perigynous. K(5), C(5), infolded in bud, A5, G(5), lobed; ovules few, axile; style 1, divided at the apex into 2 stigmas. Schizocarp. *Chile, Peru.*

A family of 1 or 2 genera, often included in the Solanaceae (as by Brummitt); there are about 16 species, a few of which are cultivated.

Genera included: ***Nolana***.

197 (229) Solanaceae. Woody/herbaceous, with internal phloem. Leaves alternate, simple/rarely pinnatisect, exstipulate; ptyxis conduplicate. Inflorescences often cymose/flowers solitary, often extra-axillary. Flowers bisexual, actinomorphic/zygomorphic. K hypogynous, CA perigynous. K5/(5), C(5), lobes folded/contorted/valvate, A5/rarely 4/2, anthers usually opening by slits/rarely by pores, G usually (2), septum usually oblique to the median plane of the flower, rarely with secondary septa; ovules *n*, axile; style 1, stigma bilobed or capitate. Berry/capsule. *Widespread.*

A large and diverse family with 90 genera and 2600 species. Fourteen genera (5 of them introduced) are found in Europe and 33 in North America. Species of about 32 genera are cultivated, either as ornamentals or as food, e.g. aubergine (*Solanum melongena*), potato (*S. tuberosum*), tomato (*Lycopersicon esculentum*), etc.

Genera included: ***Acnistus, Anisodus, Anthocercis, Atropa, Browallia, Brugmansia, Brunfelsia, Capsicum, Cestrum, Chamaesaracha, Cyphanthera, Cyphomandra, Datura, Fabiana, Hunzikeria, Hyoscyamus, Iochroma, Jaborosa, Jaltomata, Juanulloa, Leucophysalis, Lycianthes, Lycium, Lycopersicon, Mandragora, Margaranthus, Nectouxia, Nicandra, Nicotiana, Nierembergia, Nothocestrum, Oryctes, Petunia, Physalis, Physochlaina, Quincula, Salpichroa, Salpiglossis, Schizanthus, Scopolia, Solandra, Solanum, Streptosolen, Triguera, Vestia, Withania***.

198 (230) Buddlejaceae. Woody/rarely herbaceous, without internal phloem, often with glandular hairs. Leaves opposite/ whorled/rarely alternate, often toothed, stipules forming a line uniting the leaf-bases; ptyxis flat-conduplicate. Inflorescences various. Flowers bisexual, actinomorphic. K hypogynous, CA perigynous. K(4), C(4), A4, G(2), style 1, stigma capitate or 2-lobed; ovules *n*, axile. Capsule/berry/drupe. *Mainly eastern Asia.*

About 8 genera with 100 species; *Buddleja* is now introduced into both Europe and North America. The whole family is sometimes included in the Loganiaceae (see above).

Genera included: ***Buddleja, Emorya, Polypremnum.***

199 (231) Scrophulariaceae. Herbs/woody, some half-parasitic; internal phloem absent. Leaves alternate/opposite, simple, rarely compound, exstipulate; ptyxis variable. Inflorescences various. Flowers bisexual, zygomorphic. K hypogynous, CA perigynous. K(4–5), C(4–5)/rarely (–8), A4/2/rarely 5, G(2) septum at right angles to the median plane of the flower; ovules 1–*n*, axile; style 1; stigma capitate or bilobed. Capsule/rarely berry/indehiscent; seeds usually unwinged. *Widespread.*

There are over 220 genera and 4500 species. Thirty-nine genera are native to Europe and 69 to North America. Species of about 50 genera are cultivated as ornamentals.

Genera included: ***Agalinis, Alonsoa, Anarrhinum, Angelonia, Antirrhinum, Asarina, Aureolaria, Bacopa, Bartsia, Bellardia, Besseya, Boschniakia, Bowkeria, Brachystigma, Buchnera, Calceolaria, Capraria, Castilleja, Chaenorhinum, Chelone, Chionohebe, Chionophila, Collinsia, Conopholis, Cordylanthus, Cymbalaria, Cymbaria, Dermatobotrys, Derwentia, Diascia, Digitalis, Dodartia, Dopatrium, Epifagus, Epixiphium, Erinus, Euphrasia, Freylinia, Galvezia, Gratiola, Hebe, Hebenstretia, Hemiphragma, Isoplexis, Jovellana, Keckiella, Kickxia, Lafuentea, Lagotis, Leucophyllum, Limnophylla, Limosella, Linaria, Lindenbergia,***

Lindernia, Lophospermum, Macranthera, Maurandella, Maurandya, Mazus, Melampyrum, Melasma, Micranthemum, Mimetanthe, Mimulus, Misopates, Mohavea, Nemesia, Neogaerrhinum, Nothochelone, Nuttallanthus, Odontites, Orthocarpus, Ourisia, Paederota, Parahebe, Parentucellia, Pedicularis, Penstemon, Phygelius, Picrorhiza, Rehmannia, Rhinanthus, Rhodochiton, Rhynchocorys, Russellia, Sairocarpus, Schistophragma, Schlegelia, Schwalbea, Scoparia, Scrophularia, Selago, Seymeria, Sibthorphia, Siphonostegia, Striga, Sutera, Synthyris, Tetranema, Tonella, Torenia, Tozzia, Verbascum, Veronica, Veronicastrum, Wulfenia, Zaluzianskya.

200 (232) Globulariaceae. Herbs/shrublets. Leaves alternate/basal, exstipulate; ptyxis conduplicate. Inflorescence a bracteate head. Flowers bisexual, zygomorphic. K hypogynous, CA perigynous. K(5), C(4–5), A4, G(2), 1-celled, ovule 1, apical; style 1, stigma capitate. Nut. *Mostly Mediterranean area.*

A family considered here to contain only a single genus, but some authors include other, mainly South African genera (*Hebenstretia, Selago*, etc.) which are here treated as belonging to the Scrophulariaceae.

Genera included: ***Globularia.***

201 (233) Bignoniaceae. Usually woody and climbing, often with tendrils, rarely herbs. Leaves usually opposite, compound/rarely simple; ptyxis conduplicate. Inflorescence usually cymose. Flowers bisexual, zygomorphic. K hypogynous, CA perigynous. K(5), C(5), A4/rarely 2, G(2); ovules *n*, axile/rarely parietal; style 1, stigma capitate or bilobed. Fruit usually a capsule. Seeds often winged. *Mainly tropics.*

The distinction between this family and Scrophulariaceae is often difficult. There are 112 genera and 725 species; 18 genera are native to North America. Species of 28 genera are grown as spectacular ornamental climbers or flowering trees.

Genera included: ***Amphilophium, Amphitechna, Anemopaegma, Argylia, Arrabidaea, Bignonia, Campsidium, Campsis, Catalpa, Chilopsis, Clytostoma, Crescentia, Distictis, Eccremocarpus, Incarvillea, Jacaranda, Kigelia, Macfadyena, Markhamia, Pandorea, Paulownia, Podranea, Pyrostegia, Spathodea, Tabebuia, Tecoma, Tecomanthe, Tynnanthus.***

202 (234) Acanthaceae. Usually herbs. Leaves opposite, simple, exstipulate, often with cystoliths; ptyxis variable. Inflorescences cymose, often with conspicuous, overlapping bracts. Flowers bisexual, zygomorphic. K hypogynous, CA perigynous. K(4–5), C(5), 2-lipped, A4/2, G(2); ovules axile, 2 or more per cell; style 1, slightly lobed at apex. Fruit usually a capsule, which expels the seeds by force. *Mainly tropics.*

A large family with over 240 genera and about 4350 species. Two genera (1 introduced) are found in Europe, and 24 are native to North America. Species of about 40 genera are cultivated, mostly as glasshouse ornamentals, but species of *Acanthus* are generally hardy in Europe.

Genera included: ***Acanthus, Andrographis, Anisacanthus, Aphelandra, Asystasia, Barleria, Blechum, Carlowrightia, Crossandra, Dicliptera, Dyschoriste, Elytraria, Eranthemum, Fittonia, Habracanthus, Hygrophila, Hypoestes, Justicia, Mackaya, Nelsonia, Odontonema, Oplonia, Pachystachys, Peristrophe, Phialacanthus, Pseuderanthemum, Ruellia, Ruttya, Sanchezia, Siphonoglossa, Stenandrium, Strobilanthes, Teliostachya, Thunbergia, Yeatesia.***

203 (235) Pedaliaceae. Herbs, often sticky-hairy. Leaves opposite/alternate above, simple, exstipulate. Inflorescences racemes/axillary cymes/flowers solitary. Flowers bisexual, zygomorphic. K hypogynous, CA perigynous. K(5), C(5), A4 rarely 2, G(2), 2–4-celled; ovules 1–*n* per cell, axile; style 1, stigma bilobed. Capsule, often 2-horned/nut-like. *Tropics, South Africa.*

There are 18 genera and 55 species. Two genera are native to North America; *Sesamum indicum* is grown as an ornamental and produces sesame seeds, used in cooking and oil-production.

Genera included: ***Ceratotheca, Harpagophytum, Sesamum.***

204 (236) Martyniaceae. Herbs/rarely shrubby. Leaves opposite/sometimes alternate above, exstipulate. Flowers in cymes/solitary, bisexual, zygomorphic. K hypogynous, CA perigynous. K(5), C(5), 2-lipped, A4+1 staminode, G(2), 2–4-celled; ovules *n* per cell, axile; style 1, stigma bilobed. Capsule, often horned/drupe. *Subtropical America.*

There are 5 genera and about 18 species; they are sometimes included in the *Pedaliaceae* (as by Brummitt). Four of the genera are native to North America, and a few are grown as ornamentals.

Genera included: ***Martynia, Proboscidea.***

205 (237) Gesneriaceae. Herbs/shrubs, some epiphytic. Leaves usually opposite/basal, often velvety; ptyxis variable but mainly involute. Inflorescences cymose/flowers solitary. Flowers bisexual, usually zygomorphic. K hypogynous, CA perigynous/KCA epigynous. K(5), C(5), A4/2, rarely 5, G(2); ovules *n*, parietal, placentas intrusive, bifid (placentation thus appearing superficially axile); style 1, stigma capitate or slightly bilobed. Capsule/berry. *Mostly tropics.*

A large family of 146 genera and about 2400 species. Three genera are native to Europe, 5 to North America. About 46 genera are cultivated as warm-house ornamentals.

Genera included: ***Achimenes, Aeschynanthus, Alloplectus, Asteranthera, Briggsia, Chirita, Chrysothemis, Codonanthe, Columnea, Conandron, Corallodiscus, Corytoplectus, Diastema, Didymocarpus, Drymonia, Episcia, Eucodonia,***

Gesneria, Gloxinia, Haberlea, Jancaea, Koellikeria, Kohleria, Lysionotus, Mitraria, Moussonia, Nautilocalyx, Nematanthus, Neomortonia, Opithandra, Petrocosmea, Ramonda, Rhabdothamnus, Rhytidophyllum, Saintpaulia, Sarmienta, Sinningia, Smithiantha, Streptocarpus, Titanotrichum.

206 (238) Orobanchaceae. Parasitic herbs, chlorophyll 0. Leaves scale-like, alternate. Flowers bisexual, zygomorphic, K hypogynous, CA perigynous. K(4–5), C(5), A4, G(2/rarely 3); ovules *n*, parietal, usually on 4 placentas; style 1, stigma 2–3-lobed. *Mainly north temperate areas.*

There are about 15 genera and 150 species. Four genera occur in Europe and 4 are native to North America.The family is often placed in the Scrophulariaceae (as by Brummitt).

Genera included: ***Cistanche, Lathraea, Orobanche, Phelypaea.***

207 (239) Lentibulariaceae. Herbs, insectivorous, some aquatic. Leaves alternate/basal, often of 2 forms, elaborated, sometimes converted into active traps. Inflorescences on scapes, racemes/flowers solitary. Flowers bisexual, zygomorphic. K hypogynous, CA more or less perigynous. K2–5/(2–5), C(5), spurred at base, A2, G(2); ovules *n*, free-central; stigma sessile, 2-lobed or with 1 lobe sometimes reduced. Capsule. *Widespread.*

Four genera and 245 species, which trap insects either on sticky hairs or by active traps. Two genera occur in Europe, 2 in North America, and a few are grown as interesting ornamentals.

Genera included: ***Pinguicula, Utricularia.***

208 (240) Myoporaceae. Woody. Leaves usually alternate, often with resinous glands, exstipulate. Inflorescences various. Flowers bisexual, usually zygomorphic. K hypogynous, CA perigynous. K(5), C(5), A4/rarely 5, G(2); ovules 4–8, axile; style 1, stigma capitate. Fruit drupe-like. *Scattered, mostly Australia.*

A small family of 5 genera and about 220 species, a few of which are cultivated for ornament.

Genera included: ***Bontia, Myoporum.***

209 (241) Phrymaceae. Herbs. Leaves opposite, exstipulate. Inflorescence a spike. Flowers bisexual, zygomorphic, deflexed in fruit. K hypogynous, CA perigynous. K(5), teeth hooked, C(5), A4, G(2); ovule 1, basal; style 1, stigma slightly 2-lobed. Nut in persistent K. *Eastern Asia, Atlantic North America.*

A single genus with 1 or 2 species, one of which is native in North America.

Genera included: ***Phryma.***

210 (242) Plantaginaceae. Herbs/rarely shrublets. Leaves alternate/opposite/all basal. Inflorescence usually a spike. Flowers unisexual/bisexual, actinomorphic. K hypogynous, CA perigynous/KCA hypogynous. K4/(4), C(3–4), A usually 4, projecting, G1–4-celled; ovules few, axile; style 1, undivided. Capsule opening by lid/nut.

Three genera and about 250 species. Two genera are native to both Europe and North America; a few species are grown as ornamentals, but many more occur as weeds.

Genera included: ***Littorella, Plantago.***

211 (243) Caprifoliaceae. Mostly shrubs/climbers. Leaves opposite, usually simple/rarely pinnate, usually stipulate; ptyxis variable. Inflorescence often cymose. Flowers bisexual, actinomorphic/zygomorphic, often twinned, KCA epigynous. K5/(5), C usually (5), A4–5 borne on C-tube, G(3–5), sometimes only 1 cell fertile; ovules 1–*n* per cell, axile/pendulous; style short or long, stigma 3–5/lobed/sessile, free. Berry. *Widespread, mainly north temperate areas.*

Once thought to be related to the Rubiaceae (see above), and consisting of 16 genera and about 400 species. Six genera are native to Europe and 8 to North America. Species of about 10 genera are cultivated for ornament. *Sambucus* is sometimes separated off as the Sambucaceae and *Viburnum* as the Viburnaceae.

Genera included: ***Abelia, Diervilla, Dipelta, Heptacodium, Kolkwitzia, Leycesteria, Linnaea, Lonicera, Sambucus, Symphoricarpos, Triosteum, Viburnum, Weigela.***

212 (244) Adoxaceae. Rhizomatous herbs. Leaves opposite/ basal, compound, exstipulate. Flowers in a head, bisexual, actinomorphic. KCA more or less epigynous. K(2–3), C(4–6), A4–6 borne on C-tube, each split into 2 half-anthered parts, G(3–5); ovules 3–5, axile; styles 3–5, short, free. Drupe. *North temperate areas.*

A single genus (sometimes divided into 2 or more) with 2–3 species. It is native in both Europe and North America.

Genera included: ***Adoxa.***

213 (245) Valerianaceae. Herbs. Leaves opposite, simple, dissected, exstipulate; ptyxis conduplicate. Inflorescence cymose. Flowers bisexual, zygomorphic. KCA epigynous. K late-developing, sometimes pappus-like, C(5), often saccate/spurred, A1–4 borne on C-tube, G(3), only 1 cell fertile; ovule 1, pendulous; style 1, stigma 2–3-lobed. Cypsela, K often elaborated in fruit. *Mainly north temperate areas, Andes.*

Seventeen genera and about 400 species. Five genera occur in Europe and 4 in North America, and species of about 6 are cultivated.

Genera included: ***Centranthus, Fedia, Patrinia, Plectritis, Valeriana, Valerianella.***

214 (246) Dipsacaceae. Mostly herbs. Leaves opposite, simple/ dissected, exstipulate; ptyxis mostly conduplicate/rarely involute. Inflorescence an involucrate head. Flowers bisexual, zygomorphic. KCA epigynous with cupular involucel. K5–10, cupular, C(4–5), A4/rarely 2, borne on C-tube, G(2); ovule 1, apical; style

1, often with 2 small stigmatic lobes at apex. *Old World, centred in Mediterranean area.*

There are 10 genera and about 250 species. All the genera occur in Europe and 6 are introduced into North America. A few are grown as ornamentals.

Genera included: ***Cephalaria, Dipsacus, Knautia, Pterocephalus, Scabiosa, Succisa, Succisella.***

214a (247) Morinaceae. Herbs. Leaves in whorls, mostly basal, evergreen, spiny-margined, exstipulate. Inflorescence a spike of many-flowered, bracteate whorls, which may sometimes coalesce. Flowers bisexual, zygomorphic. KCA epigynous. K(2), within bristle-tipped involucel, C(5), tube curved, 2-lipped, A2, G(2); ovule 1, apical; style 1. *Europe east to the Himalaya.*

A single genus native to Europe and occasionally cultivated.

Genera included: ***Morina.***

215 (248) Campanulaceae. Herbs, often with milky sap/rarely woody. Leaves usually alternate, simple, exstipulate; ptyxis variable but mainly supervolute. Inflorescence various. Flowers bisexual, actinomorphic/zygomorphic. KCA epigynous/rarely hypogynous. K5/rarely 3–10, C(5)/rarely (3–10), valvate, A5/rarely 3–10, rarely borne on C-tube, G(2–5)/rarely (–10); ovules *n*, axile; style 1, stigma shallowly or deeply 2–5-lobed. Capsule/fleshy. *Widespread.*

A family of 87 genera and nearly 2000 species. Fifteen genera are native to Europe, 23 to North America, and species of about 20 genera are grown as ornamentals, The genera with zygomorphic corollas are sometimes split off as the separate family Lobeliaceae.

Genera included: ***Adenophora, Asyneuma, Azorina, Brighamia, Campanula, Canarina, Clermontia, Codonopsis, Cyananthus, Diosphaera, Downingia, Edraianthus, Githopsis, Hanabusaya, Heterocodon, Howellia, Jasione, Laurentia, Legenere, Legousia, Leptocodon, Lobelia, Michauxia, Monopsis, Musschia, Nemacladus, Ostrowskia, Parishella, Petromarula,***

Physoplexis, Phyteuma, Platycodon, Pratia, Sphenoclea, Symphyandra, Trachelium, Triodanis, Wahlenbergia.

216 (249) Goodeniaceae. Herbs/shrubs. Leaves usually alternate, simple, exstipulate. Flowers bisexual, zygomorphic. KCA usually epigynous. K5/(5), C(5), 1–2-lipped, lobes valvate/infolded, A5, sometimes borne on C-tube, G(2), 1–2-celled; ovules 1–2 per cell, axile/basal; style 1, stigmas 2, sheathed. Fruits various. *Mainly Australia.*

Sixteen genera and 430 species. One genus is native to North America. Several genera have recently become popular as garden ornamentals.

Genera included: ***Dampiera, Goodenia, Leschenaultia, Scaevola, Selliera.***

217 (250) Brunoniaceae. Herbs. Leaves basal, simple, exstipulate. Inflorescence a head with bracts. Flowers bisexual, more or less actinomorphic. K hypogynous, CA perigynous. K(5), C(5), valvate, A(5), the anthers united into a tube around the style, G1-celled; ovule 1, basal; style 1, stigma sheathed. Nut enclosed in K-tube. *Australia.*

A single genus with a single species, occasionally grown as an ornamental in Europe. Often included in the Goodeniaceae.

Genera included: ***Brunonia.***

218 (251) Stylidiaceae. Herbs. Leaves basal/alternate, linear, usually exstipulate. Inflorescences various. Flowers unisexual/bisexual, actinomorphic/zygomorphic. KCA epigynous. K(5–7), C(5), imbricate, A2 joined to style, G(2); ovules *n*, axile/parietal/free central; style 1, with stamens attached, held to 1 side of the flower, moving rapidly centrally when touched at the base. Fruit usually a capsule. *Australasia.*

Three genera with 270 species; a few grown as curiosities on account of their touch-sensitive styles.

Genera included: ***Stylidium.***

218a (252) Calyceraceae. Herbs. Leaves alternate, exstipulate. Flowers bisexual, actinomorphic/zygomorphic, in an involucrate head. KCA epigynous. K4–6, C(4–6), A4–6, arising near mouth of corolla-tube, filaments often united at base, G apparently 1 with a single apical ovule. Fruit an achene. *Mostly Central and South America.*

Genera included: ***Acicarpha, Nastanthus.***

219 (253) Compositae/Asteraceae. Herbs/woody, sometimes with milky sap. Leaves variable, alternate/opposite/whorled, exstipulate but occasionally with small leaflets at the base resembling stipules; ptyxis variable. Inflorescence an involucrate head (rarely heads 1-flowered and aggregated into second order heads). Flowers unisexual/bisexual, actinomorphic/zygomorphic. KCA epigynous. K reduced to pappus/scales/rarely completely absent, C(5/3). A5/rarely (3), anthers united into a tube around the style, G(2); ovule 1, basal; style 1, stigma 2-lobed (the stigmatic lobes may sometimes be longer than the style). Cypsela, usually with pappus. *Widespread.*

The largest family of dicotyledons, with about 1300 genera and 21 000 species. One hundred and eighty-one genera are native to Europe, 346 to North America. Many arre grown as ornamentals, vegetables and flavourings.

Genera included: ***Acamptopappus, Acanthospermum, Achillea, Achyrachaenia, Acourtia, Adenocaulon, Adenostemma, Aetheorhiza, Ageratum, Agoseris, Ajania, Allardia, Amberboa, Amblyoplepis, Ambrosia, Ammobium, Amphiachyris, Amphipappus, Amphoricarpos, Anacyclus, Anaphalis, Andryala, Anisocoma, Antennaria, Anthemis, Antheropeas, Aphanostephus, Aposeris, Arctanthemum, Arctium, Arctotheca, Arctotis, Argyranthemum, Argyroxiphium, Arnica, Arnoseris, Artemisia, Aster, Asteriscus, Astranthium, Atractylis, Atrichoseris, Baccharis, Bahia, Baileya, Balduina, Balsamorhiza, Bartlettia, Bebbia, Bedfordia, Bellis, Bellium, Benitoa, Berardia, Berkheya, Berlandiera, Bidens, Blennosperma,***

Blepharipappus, Blepharizonia, Blumea, Boltonia, Bombycilaena, Borrichia, Brachyglottis, Bracteantha, Bradburia, Brickellia, Buphthalmum, Cacalia, Cacaliopsis, Calendula, Callistephus, Calomeria, Calycadenia, Calycocorsus, Calycoseris, Calyptocarpus, Cardopatium, Carduncellus, Carduus, Carlina, Carminatia, Carpesium, Carpephorus, Carthamus, Cassinia, Catananche, Celmisia, Centaurea, Centipeda, Cephalorrhynchus, Chaenactis, Chaetadelpha, Chaetopappa, Chamaemelum, Chaptalia, Cheirolophus, Chiliotrichium, Chlamydophora, Chondrilla, Chrysactinia, Chrysanthemoides, Chrysanthemum, Chrysocoma, Chrysogonum, Chrysopsis, Chrysothamnus, Cicerbita, Cichorium, Cirsium, Cladanthus, Clappia, Clibadium, Cnicus, Coleostephus, Conyza, Coreocarpus, Coreopsis, Corethrogyne, Cosmos, Cotula, Cousinia, Craspedia, Cremanthodium, Crepis, Crupina, Cynara, Dahlia, Daveaua, Dendranthema, Dichaetophora, Dichrocephala, Dicoria, Dicranocarpus, Dimeresia, Dimorphotheca, Dittrichia, Doronicum, Dubautia, Dugaldia, Dyssodia, Eastwoodia, Eatonella, Echinacea, Echinops, Eclipta, Egletes, Elephantopus, Eleutheranthera, Emilia, Encelia, Enceliopsis, Engelmannia, Enydra, Erechtites, Ericameria, Erigeron, Eriocephalus, Eriophyllum, Eumorphia, Eupatorium, Euryops, Euthamia, Evacidium, Facelis, Felicia, Filago, Flaveria, Florestina, Flourensia, Gaillardia, Galactites, Galinsoga, Gamochaeta, Garberia, Gazania, Geraea, Gerbera, Glossopappus, Glyptopleura, Gnaphalium, Gochnatia, Grindelia, Guardiola, Guizotia, Grundlachia, Gutierezia, Gymnosperma, Gynura, Haplocarpha, Haploesthes, Haplopappus, Hartwrightia, Hazardia, Hecastocleis, Hedypnois, Helenium, Helianthella, Helianthus, Helichrysum, Heliopsis, Hemizonia, Hesperomannia, Heteracia, Heteranthemis, Heterosperma, Heterotheca, Hieracium, Hippia, Hispidella, Hofmeisteria, Holocarpha, Holozonia, Homogyne, Hulsea, Hymenoclea, Hymenonema, Hymenopappus, Hymenostemma, Hymenothrix, Hymenoxys, Hyoseris, Hypochaeris, Ifloga, Inula, Ismelia,

Isocarpha, Iva, Ixeris, Jamesianthus, Jasonia, Jurinea, Kalimeris, Karelinia, Keysseria, Kleinia, Koelpinia, Krigia, Lactuca, Lagascea, Lagenophora, Lagophylla, Lamyropsis, Lapsana, Lasiopogon, Lasthenia, Launaea, Layia, Leibnitzia, Leontodon, Leontopodium, Lepidophorum, Lepidospartum, Leptinella, Leptocarpha, Lessingia, Leucanthemella, Leucanthemopsis, Leucanthemum, Leucogenes, Leuzea, Leysera, Liatris, Ligularia, Lindheimera, Lipochaeta, Logfia, Lonas, Luina, Lygodesmia, Machaeranthera, Madia, Malacothrix, Malperia, Mantisalca, Marshallia, Matricaria, Megalodonta, Melampodium, Melanthera, Micropus, Microseris, Mikania, Monolopia, Monoptilon, Montanoa, Munzothamnus, Mutisia, Mycelis, Nananthea, Nardophyllum, Nassauvia, Neurolaena, Nicolletia, Nipponanthemum, Nolletia, Nothocalais, Notobasis, Olearia, Omalotheca, Onopordum, Orochaenactis, Osteospermum, Otanthus, Othonna, Otospermum, Oxylobus, Ozothamnus, Palaeocyanus, Palafoxia, Parthenice, Parthenium, Pectis, Pentachaeta, Pentzia, Perezia, Pericallis, Pericome, Perityle, Petasites, Petradoria, Peucephyllum, Phagnalon, Phalacrachena, Phalacrocarpum, Phalacroseris, Phoebanthus, Picnomon, Picris, Pilosella, Pinaropappus, Piptocarpha, Piptocoma, Piqueria, Pityopsis, Plagius, Platyschkuhria, Pluchea, Plummera, Porophyllum, Prenanthella, Prenanthes, Prionopsis, Prolongoa, Proustia, Psathyrotes, Pseudobahia, Pseudoclappia, Psilocarphus, Psilotrophe, Pterocaulon, Ptilostemon, Pulicaria, Pyrrhopappus, Rafinesquia, Raillardia, Rainiera, Raoulia, Ratibida, Reichardia, Remya, Rhagadiolus, Rhodanthe, Rhodanthemum, Rolandra, Rothmaleria, Rudbeckia, Sachsia, Salmea, Santolina, Sanvitalia, Sartwellia, Saussurea, Schkuhria, Sclerocarpus, Sclerolepis, Scolymus, Scorzonera, Senecio, Serratula, Shinneroseris, Sigesbeckia, Silphium, Silybum, Simsia, Sinacalia, Solidago, Soliva, Sonchus, Sphaeromeria, Spilanthes, Staehelina, Stenotus, Stephanomeria, Stevia, Stokesia, Struchium, Stylocline,

Syncarpha, Synedrella, Syntrichopappus, Tagetes, Tanacetum, Taraxacum, Telekia, Tessaria, Tetradymia, Tetragonotheca, Tetramolopium, Thelesperma, Thurovia, Thymophylla, Tithonia, Tolpis, Townsendia, Tracyina, Tragopogon, Trichocoronis, Trichoptilium, Tridax, Trixis, Tussilago, Tyrimnus, Urospermum, Ursinia, Vanclevea, Varilla, Venegasia, Verbesina, Vernonia, Viguiera, Volutaria, Wagenitzia, Wedelia, Whitneya, Wilkesia, Wyethia, Xanthisma, Xanthium, Xanthocephalum, Xeranthemum, Youngia, Zaluzania, Zinnia.

Subclass Monocotyledones

Cotyledon 1, terminal; leaves usually with parallel veins, sometimes these connected by cross-veinlets; leaves without stipules, opposite only in some aquatic plants; flowers usually with parts in 3s; mature root-system wholly adventitious, the primary root soon dying away.

220 (254) Alismataceae. Aquatic herbs, often scapose, without latex. Leaves alternate, often broad; ptyxis supervolute. Inflorescence usually much-branched, rarely a single umbel. Flowers unisexual/ bisexual, actinomorphic. Perianth 2-whorled: K3, C3, A6–*n*, G6–*n*, superior; ovules 1–2 per carpel, basal/marginal. Fruit a group of achenes. *Temperate and tropical areas.*

Thirteen genera and about 90 species of water plants. Six genera are native to Europe, 4 to North America. Species of 5 genera are grown as ornamentals.

Genera included: ***Alisma, Baldellia, Caldesia, Damasonium, Echinodorus, Limnophyton, Luronium, Machaerocarpus, Sagittaria.***

221 (255) Butomaceae. Aquatic herbs, without latex. Leaves basal, not clearly divided into blade and stalk; ptyxis supervolute.

Flowers solitary or in umbels, subtended by bracts. Flowers bisexual, actinomorphic. Perianth 1-whorled, all segments petal-like; P6, A1–*n*, G6–*n*, superior; ovules many in each carpel, placentation diffuse-parietal. Fruit a follicle or group of follicles. *Temperate Eurasia.*

A single genus, native to Europe, with a single species, often cultivated as an ornamental aquatic.

Genera included: ***Butomus.***

221a (256) Limnocharitaceae. Aquatic herbs, with latex. Leaves on the stems, clearly divided into a blade and stalk. Flowers solitary or in umbels subtended by bracts. Flowers bisexual, actinomorphic. Perianth of 2 whorls: K3, C3, persistent, A1–*n*, G6–*n*, superior; ovules many in each carpel, placentation diffuse-parietal. Fruit a follicle or group of follicles. *Tropics.*

Three genera, species of 2 of them cultivated as ornamental aquatics.

Genera included: ***Hydrocleys, Limnocharis.***

222 (257) Hydrocharitaceae. Aquatics, usually with at least the flowers emerging from the water. Leaves alternate, variable, usually with distinct stalk and blade. Flowers usually borne in a bifid spathe or between 2 opposite bracts, rarely solitary, unisexual or bisexual, actinomorphic. Perianth in 2 whorls: K3, C3, A1–*n*, G usually (3–6), inferior; ovules many, diffuse parietal. Fruit usually a capsule, rarely berry-like. *Mainly in tropical and warm temperate areas.*

There are 15 genera and about 100 species of submerged, floating or emergent aquatics. Ten genera occur wild in Europe, and 10 in North America. Species of 9 genera are grown as ornamentals or aquarium-oxygenating plants.

Genera included: ***Blyxa, Egeria, Elodea, Halophila, Hydrilla, Hydrocharis, Lagarosiphon, Limnobium, Ottelia, Stratiotes, Thalassia, Vallisneria.***

223 (258) Scheuchzeriaceae. Herbaceous bog plants, sometimes more or less aquatic. Leaves in 2 ranks, with sheathing bases each bearing a ligule. Flowers in racemes, with bracts, bisexual, actinomorphic. Perianth 1-whorled: P6, A6, G3/6, superior; ovules 2 per carpel, basal. Fruit a group of follicles. *Cold north temperate areas.*

A single genus with a single species (*Scheuchzeria palustris*) native to both Europe and North America.

Genera included: ***Scheuchzeria***.

224 (259) Aponogetonaceae. Aquatics. Leaves alternate, long-stalked, with sheathing bases and expanded blades. Inflorescence a simple or forked spike. Flowers usually bisexual and actinomorphic. Perianth 1-whorled: P1–3/rarely absent/or, when P1, bract-like, A6 or more, G3–6, superior; ovules few in each carpel, basal. Fruit a group of follicles. *Mainly Old World tropics.*

A single genus with about 44 species, a few of which are cultivated as ornamental aquatics.

Genera included: ***Aponogeton***.

225 (260) Juncaginaceae. Usually marsh plants, sometimes halophytic. Leaves all basal, sheathing. Inflorescence a raceme/spike, without bracts. Flowers unisexual/bisexual, actinomorphic. Perianth 1-whorled: P6 or rarely 1 (when sometimes interpreted as a bract), A1/4–6, G(3–6), superior, 1-celled; ovules 1 per cell, basal. Fruit a capsule or indehiscent and of 2 forms. *Mainly temperate and cold regions; Pacific North America.*

Four genera and 18 species, often found in brackish water. One genus is native to Europe, 1 to North America.

Genera included: ***Triglochin***.

225a (261) Lilaeaceae. Aquatic or marsh herbs. Leaves all at the base of the scapes, with open sheaths at the base and blades thick, linear and spongy. Flowers in inflorescences of 2 kinds, unisexual; male flowers in an axillary spike; P1, A1; female flowers P1, G1, ovule 1, basal: some at the bases of the male spikes, short-styled, others sessile within the leaf-sheaths and with very long, thread-like styles. Fruits dimorphic, those from the spikes narrowly winged and shortly beaked, those from the basal flowers 3-angled and unequally 3-horned at the apex; all indehiscent. *Western North America from British Columbia to Chile and Argentina.*

A single genus with a single species.

Genera included: ***Lilaea.***

226 (262) Potamogetonaceae. Submerged or emergent aquatics of fresh or brackish water. Leaves alternate or opposite, sometimes in 2 ranks, sometimes with ligules, or stipule-like sheath-margins. Inflorescence a spike. Flowers unisexual/bisexual, actinomorphic. Perianth 1-whorled: P4, segments clawed, valvate, A4, G4, free, superior. Fruit indehiscent. *Widespread.*

Two genera with over 100 species, both native to Europe and North America. A few are grown mainly as aquarium plants.

Genera included: ***Groenlandia, Potamogeton.***

226a (263) Ruppiaceae. Herbaceous plants of brackish marshes. Leaves opposite, with sheathing bases but without ligules. Flowers bisexual, actinomorphic, sessile in 2-flowered spikes at first enclosed between the bases of the leaves. P0, A2 G4; ovule 1 per carpel, pendulous. Fruiting carpels long-stalked, indehiscent. *Tropical and temperate areas.*

A single genus with about 7 species; it is native to Europe.

Genera included: ***Ruppia.***

226b (264) Zosteraceae. Submerged marine perennials with rhizomes. Leaves in 2 ranks, sheathing at the base, sheaths with stipule-like margins. Flowers arranged in rows on a flattened axis, consisting of alternating stamens and carpels, each opposite pair (considered as forming a bisexual flower) being covered by a bract in bud. P0, A consisting of 2 sessile half-anthers, G(2), superior, 1-celled; ovule 1, pendulous. Fruit indehiscent. *Widespread.*

Three genera and 17 species; one genus is native to Europe, 2 to North America.

Genera included: ***Phyllospadix, Zostera.***

226c (265) Posidoniaceae. Submerged marine perennials with densely fibrous rhizomes (often washed up on beaches as fibre-balls). Leaves mostly basal, with ligules. Flowers bisexual, in spikes. P4, A4, borne at the bases of the perianth-segments, G1-celled, superior, stigma sessile; ovule 1, parietal. Fruit fleshy. *Mediterranean area, Australia.*

There is a single genus (native to Europe) with perhaps 3 species.

Genera included: ***Posidonia.***

227 (266) Zannichelliaceae. Submerged aquatics of fresh or salty water. Leaves alternate/opposite/whorled, entire. Flowers unisexual in axillary cymes/solitary. Perianth cupular/of 3 scales/absent, A1–3, G1–9, superior/naked, stigmas dilated above or 2–4-lobed; ovule 1 per carpel, pendulous. Fruit stalked, indehiscent. *Widespread.*

Four genera with 7 species. There are 2 genera native to both Europe and North America.

Genera included: ***Althenia, Zannichellia.***

227a (267) Cymodoceaceae. Marine herbs with rhizomes. Leaves mostly alternate, linear, each with a ligule. Flowers axillary, solitary or in cymes. P absent, A 2, G(2–3) with a single apical ovule. Fruit a nutlet. *Warm coastal shallows.*

Genera included: ***Cymodocea, Halodule.***

228 (268) Najadaceae. Submerged aquatics of fresh or salty water. Leaves opposite/whorled, entire/toothed. Flowers unisexual in axillary cymes/solitary. Male: P2-lipped, A1; female P membranous/absent, G1-celled, superior/naked, with 2–4 stigmas; ovule 1, basal. Fruit indehiscent. *Widespread.*

A single genus with 11 species; it is native to both Europe and North America.

Genera included: ***Najas***.

229 (269) Liliaceae. Plants with bulbs, rhizomes or corms, terrestrial, herbaceous. Leaves generally borne on the stems. Flowers in racemes/panicles, without spathes when solitary. P6/(6), well-developed; nectaries on the perianth-segments, rarely absent, A6, G(3), superior, ovules *n*, axile; style 1 sometimes divided above or stigmas sessile. Capsule. *Mainly north temperate areas.*

In the restricted usage employed here, there are 17 genera and about 415 species. Five genera are native to Europe, 6 to North America. The 'old' Liliaceae was a much more inclusive concept, accommodating the genera treated here in the families numbered 226a–226s (amounting to about 290 in total, with 4500 species). The assignment of genera to these smaller families is still somewhat controversial; here, 3 genera more often included in the Colchicaceae (*Gloriosa, Littonia* and *Sandersonia*) are put with the reduced Liliaceae, as are two others whose family placement is very controversial: *Tricyrtis* and *Uvularia*. All of these genera have conspicuous nectaries on the perianth, and an overall similarity to genera always retained in the Liliaceae. *Calochortus* is sometimes separated off as the Calochortaceae, *Tricyrtis* as the Tricyrtidaceae and *Uvularia* as the Uvulariaceae.

Genera included: ***Calochortus, Cardiocrinum, Erythronium, Fritillaria, Gagea, Gloriosa, Lilium, Littonia, Lloydia, Nomocharis, Notholirion, Sandersonia, Tricyrtis, Tulipa, Uvularia***.

229a (270) Melanthiaceae. Plants terrestrial, herbaceous, with rhizomes. Leaves well developed, on the stems. Flowers in racemes/panicles, if solitary, without spathes, bisexual. P6, nectaries absent or on perianth-segments, A6, G3/(3), usually superior, rarely half-inferior, not stalked; ovules numerous, placentation axile; styles 3. Capsule. Seeds not hairy. *Northern hemisphere and South America.*

There are 22 genera with about 80 species; 4 are found in Europe, 14 in North America, and several are cultivated. Includes *Nartheciaceae.*

Genera included: ***Aletris, Amianthium, Chamaelirium, Chionographis, Harperocallis, Helonias, Heloniopsis, Lophiola, Melanthium, Narthecium, Schoenocaulon, Stenanthium, Tofieldia, Veratrum, Xerophyllum, Zigadenus.***

229b (271) Asphodelaceae. Plant terrestrial, herbaceous, with rhizomes. Leaves borne on the flowering stem, not reduced to sheaths. Flowers in racemes/panicles, without spathes when solitary, bisexual. P6/(6), A6, G(3), superior, ovules many, axile; style 1; nectar secreted on the ovary. Capsule. *Old World, Africa, New Zealand.*

Twelve genera and about 150 species. Five genera are native to Europe, and a few are cultivated.

Genera included: ***Asphodeline, Asphodelus, Bulbine, Bulbinella, Eremurus, Kniphofia, Paradisea, Simethis.***

229c (272) Anthericaceae. Plants terrestrial, usually with rhizomes. Leaves all basal, without broadened blades and petiole-like bases, not fleshy and succulent, nor reduced to sheaths. Flowers in racemes, spikes or panicles, bisexual. P6, free or very slightly united at the base, all similar, A6, not declinate, G(3), superior, nectar secreted on the ovary; placentation axile. Capsule. *Widespread.*

There are 9 genera with about 230 species. One genus is native to Europe and 2 to North America; a few are cultivated for ornament.

Genera included: ***Anthericum, Arthropodium, Chlorophytum, Eremocrinum, Herpolirion, Leucocrinum, Pasithea, Thysanotus.***

229d (273) Aphyllanthaceae. Plants terrestrial, with rhizomes. Leaves without blades, reduced to basal sheaths. Flowers in heads, bisexual. P6, A6, G(3), superior, placentation axile; nectar secreted on the ovary. *West Mediterranean area.*

A single genus, native to Europe and sometimes cultivated.

Genera included: ***Aphyllanthes***.

229e (274) Hemerocallidaceae. Plant terrestrial, entirely herbaceous, usually with a rhizome. Leaves all basal, without broadened blades and petiole-like bases, not fleshy and succulent, nor reduced to basal sheaths. Flowers in racemes, spikes or panicles, bisexual, actinomorphic/slightly zygomorphic. P(6), funnel-shaped, united into a tube below, A6, declinate, G(3), superior, nectar secreted on the ovary; ovules *n*, placentation axile. Capsule. *Central Europe to Japan.*

As treated here, a family of a single genus with 15–20 species, native to Europe, many species and hybrids cultivated for ornament. Genera here treated as *Phormiaceae* (below) are sometimes combined with the *Hemerocallidaceae*.

Genera included: ***Hemerocallis***.

229f (275) Hostaceae. Plant terrestrial, entirely herbaceous, usually with rhizomes. Leaves all basal, with a broadened blade and petiole-like basal part, not fleshy and succulent. Flowers in a 1-sided raceme. P(6), united below into a narrow tube, which broadens upwards, A6, G(3), superior, placentation axile, nectar secreted on the ovary. Capsule. *Japan, Korea, China.*

A single genus with about 26 species, very widely cultivated, sometimes found in gardening literature under the synonym Funkiaceae.

Genera included: ***Hosta***.

229g (276) Blandfordiaceae. Plants terrestrial, herbaceous. Leaves well developed, mainly basal. Flowers in racemes, bisexual. P(6), nectar secreted on the perianth. A6, G(3), superior, stalked, placentation axile, style 1. Capsule. Seeds hairy. *Eastern Australia.*

A single genus with 4 species, occasionally cultivated.

Genera included: ***Blandfordia.***

229h (277) Aloaceae. Plants terrestrial, herbaceous, with rhizomes. Leaves all basal, fleshy and succulent. Flowers in racemes, bisexual. P3+3, those of the outer whorl united for part of their length, those of the inner whorl variably joined to the outer, all red, orange or yellow, tips often greenish, A6, G(3), placentation axile; nectar secreted on the ovary. Capsule. *Mainly southern Africa.*

A family of 7 genera and about 560 species. Many are cultivated as ornamental succulents.

Genera included: ***Aloe, Gasteria, ×Gastrolea, Haworthia, Poellnitzia.***

229i (278) Colchicaceae. Plants terrestrial, herbaceous, with corms. Leaves well developed. Flowers in racemes or without spathes if solitary. P6/(6), if free then often connivent, A6, nectaries on the staminal filaments. G(3), superior, not stalked, below ground-level in corm; placentation axile; styles 3. Capsule. Seeds not hairy. *Old World.*

There are 8 genera and about 140 species; several genera are cultivated. Several genera often found in this family have been removed here to the Liliaceae (see above).

Genera included: ***Androcymbium, Bulbocodium, Colchicum, Merendera.***

229j (279) Alstroemeriaceae. Terrestrial plants, herbaceous. Leaves reversed by a twist at the base, not hairy or pleated. Flowers bisexual in an umbel without spathes, sometimes with additional flowers below the terminal umbel. P6, weakly zygomorphic, segments generally similar, but often 2 or 3 of the inner whorl spotted, A6, G(3), fully inferior; placentation axile. Capsule. Seeds pale. *Central and South America.*

There are 5 genera with 65 species. One genus is native to the southern parts of the USA, and two are widely cultivated.

Genera included: ***Alstroemeria, Bomarea.***

229k (280) Hyacinthaceae. Plants terrestrial, herbaceous, with bulbs. Leaves basal, well developed. Flowers in racemes or panicles, when solitary without spathes. P6/(6), A6, G(3), superior; placentation axile; nectar secreted on the ovary. Capsule. *Widespread.*

A large family of bulbous plants with 67 genera and about 950 species. Twelve genera are native to Europe, 9 to North America, and many are grown as ornamentals.

Genera included: ***Albuca, Bellevalia, Bowiea, Brimeura, Camassia, Chionodoxa, Chlorogalum, Daubenya, Dipcadi, Drimia, Drimiopsis, Eucomis, Hastingsia, Hesperocallis, Hyacinthella, Hyacinthoides, Hyacinthus, Lachenalia, Ledebouria, Massonia, Muscari, Ornithogalum, Puschkinia, Schoenolirion, Scilla, Urginea, Veltheimia.***

229l (281) Alliaceae. Plants herbaceous, terrestrial often with both bulbs and/or rhizomes, often smelling of onion or garlic. Leaves well developed. Flowers in umbels subtended by spathes, or solitary and subtended by spathes. Perianth-segments 6/(6), A6, filaments sometimes winged, G(3), superior; placentation axile. Capsule. *Temperate areas.*

There are 13 genera and 650 species. Five genera are native to Europe, 12 to North America, and many are cultivated as ornamentals, vegetables or flavourings. *Agapanthus* is sometimes separated off as Agapanthaceae.

Genera included: ***Agapanthus, Allium, Androstephium, Bessera, Bloomeria, Brodiaea, Caloscordum, Dichelostemma, Ipheion, Leucocoryne, Milla, Muilla, Nectaroscordum, Nothoscordum, Triteleia, Triteleiopsis, Tulbaghia.***

229m (282) Convallariaceae. Plants terrestrial, entirely herbaceous, usually with rhizomes. Foliage leaves on flowering stem, with well developed blades. Flowers in racemes/clusters, without spathes when solitary. P(4/6) tubular to urceolate, A4/6, G(2/3), superior or partly inferior; placentation axile; nectar secreted from glands on the ovary. Fruit a berry. *Northern hemisphere.*

Seventeen genera and 210 species. Five genera native to Europe, 10 to North America, several cultivated for ornament.

Genera included: ***Aspidistra, Clintonia, Convallaria, Disporum, Liriope, Maianthemum, Medeola, Ophiopogon, Peliosanthes, Polygonatum, Reineckia, Rohdea, Speirantha, Streptopus.***

229n (283) Asteliaceae. Plants terrestrial, herbaceous. Leaves well developed, borne on stems, often with a surface of scales which can be peeled off. Flowers in panicles, unisexual, whitish. P6, A6, G(3); styles 3; ovules numerous, placentation axile. Capsule. *Southern hemisphere (not Africa).*

There are 4 genera with 30 species; few are cultivated.

Genera included: ***Astelia.***

229o (284) Trilliaceae. Plants with rhizomes. Leaves borne on the stems, broad, opposite or in a single whorl. Flowers bisexual, solitary or in umbels without spathes. P4/6, rarely in 2 distinct whorls, 3+3, A4/6, G(3–5), superior, placentation axile/parietal. Capsule. *Temperate northern hemisphere.*

There are 4 genera and about 70 species. One genus is native to Europe, 2 to North America, and species of 3 are cultivated as woodland ornamentals.

Genera included: ***Paris, Scoliopus, Trillium.***

229p (285) Asparagaceae. Plants herbaceous or woody. Leaves reduced to small scale-like spines on the stems, their function taken over by cladodes which are flattened or much divided. Flowers borne among the cladodes. P6, A6, G(3), superior; ovules *n*, axile. Fruit a berry. *Old World, Australia.*

One or possibly two genera, with about 300 species; one genus native to Europe.

Genera included: ***Asparagus***.

229q (286) Ruscaceae. Plants woody. Leaves reduced to scales, their function taken over by flattened, leaf-like cladodes. Flowers borne directly on cladodes in clusters, bisexual or unisexual. P6/(6), A6/3, filaments united below, G(3), superior, 1-celled. Berry. *Canary Islands to Iran.*

Three genera with 8 species, one native to Europe.

Genera included: ***Danae, Ruscus, Semele***.

229r (287) Philesiaceae. Woody climbers, shrubs or subshrubs. Leaves on the stems, well developed, leathery or parchment-like, not succulent, clearly stalked. Flowers conspicuous, usually large. P6/(6), A6, sometimes attached to the base of the perianth, G(3), superior, usually 1-celled; placentation usually parietal. Berry. *Southern hemisphere (absent from Africa).*

Seven genera and about 10 species, some cultivated for their large, attractive flowers. *Luzuriaga* is sometimes separated off as the Luzuriagaceae.

Genera included: ***Eustrephus, Geitonoplesium, Lapageria, Luzuriaga, ×Philageria, Philesia***.

229s (288) Smilacaceae. Dioecious, woody climbers/more rarely herbaceous plants. Leaves on the stems, leathery or parchment-like, not succulent, clearly stalked, stalks bearing 2 tendrils. Flowers inconspicuous. P6, A6, G(3), superior; ovules few, axile. Berry. *Mainly tropics.*

Three genera with about 320 species; one genus native to both Europe and North America.

Genera included: ***Smilax***.

230 (289) Agavaceae. Succulent herbs, often large. Leaves mostly in basal rosettes, succulent, ptyxis flat or revolute, each leaf terminating in a hard or soft spine. Flowers in racemes/panicles, often produced after prolonged vegetative growth. Flowers usually bisexual, actinomorphic/zygomorphic, less than 8 cm long. P6/(6), A6, usually borne near the top of the perianth-tube, G(3), superior/inferior; ovules *n*, axile. Berry/dry and indehiscent. *North and South America, southern hemisphere.*

There are 8 genera and perhaps 380 species. Seven genera are native to North America, and two are introduced into Europe. All are cultivated. The families listed below as 230a–d have all been included in the Agavaceae in the past.

Genera included: ***Agave, Beschorneria, Cordyline, Furcraea, Hesperaloe, Polianthes, Yucca.***

230a (290) Doryanthaceae. Plant forming a loose to dense rosette, stems not obvious. Leaves leathery, blade-shaped, each leaf terminating in a cylindrical body. Flowers 10–15 cm long, bisexual, actinomorphic to weakly zygomorphic, in panicles. P6, spreading, A6, G(3), inferior; ovules *n*, axile. Capsule. *Eastern Australia.*

A single genus with 2 species, occasionally cultivated.

Genera included: ***Doryanthes.***

230b (291) Dracaenaceae. Plants mostly woody. Leaves erect and blade-like, forming a tuft on a woody rhizome, or in terminal rosettes, not stalked, parchment-like or leathery. Flowers all bisexual, actinomorphic to weakly zygomorphic, in racemes or panicles, flower stalks not jointed. P6, A6, G(3), inferior, ovules 1–*n*, axile. Berry. *Tropics, Canary Islands.*

There are 2 genera, both commonly cultivated. *Cordyline*, currently included in the Agavaceae (above), may perhaps be better placed here.

Genera included: ***Dracaena, Sansevieria.***

230c (292) Nolinaceae. Shrubs/trees, base of the trunk often swollen. Leaves not stalked, leathery or parchment-like. Flowers in panicles, mostly unisexual; flower-stalks jointed near their bases. P6/(6), A6, G(3), inferior; ovules 2 in each cell of the ovary. Fruit dry, indehiscent. *Warm parts of America.*

Three genera with about 50 species. One is native to North America, and all are cultivated.

Genera included: ***Beaucarnea, Dasylirion, Nolina***.

230d (293) Phormiaceae. Plants terrestrial, herbaceous. Leaves equitant, distichous, forming fans. Flowers in racemes, bisexual or occasionally male by abortion. P(6), A6, G(3), usually 3-celled, placentation axile. Capsule/berry. *Asia, Australasia.*

Seven genera with about 30 species, several cultivated.

Genera included: ***Dianella, Phormium, Stypandra***.

231 (294) Haemodoraceae. Herbs, sap often orange. Leaves mostly basal, often equitant. Flowers in cymes/racemes/clusters, bisexual, actinomorphic/weakly zygomorphic. P6/(6), persistent, often densely hairy outside, A6/3, G(3), superior/inferior; ovules *n*, axile. Fruit a capsule. *Mainly southern hemisphere and North America.*

Twenty-two genera with about 160 species. Two genera are native to North America, and about 5 are cultivated.

Genera included: ***Anigozanthos, Conostylis, Lachnanthes, Macropidia, Xiphidium***.

232 (295) Amaryllidaceae. Herbs with bulbs/rhizomes/corms. Leaves basal, often in 2 ranks, ptyxis flat, rarely supervolute. Flowers solitary or in umbels, the flower or umbel subtended by 1–several spathes. Flowers bisexual, actinomorphic/zygomorphic. P6/(6), A6, anthers occasionally opening by pores, G(3), inferior; ovules 2–*n* per cell, axile. Fruit a capsule/berry. Seeds black.

There are about 70 genera with about 850 species. Seven genera are native to Europe, 9 to North America. Many are cultivated, especially from the genera *Hippeastrum, Galanthus* and *Narcissus*.

Genera included: ***Amaryllis, Brunsvigia, Caliphruria, Chlidanthus, Clivia, Crinum, Cyrtanthus, Eucharis, Galanthus, Griffinia, Habranthus, Haemanthus, Hippeastrum, Hymenocallis, Lapiedra, Leucojum, Lycoris, Narcissus, Nerine, Pamianthe, Pancratium, Paramongaia, Phaedranassa, Pyrolirion, Scadoxus, Sprekelia, Stenomesson, Sternbergia, Urceolina, Vagaria, Zephyranthes.***

232a (296) Ixioliriaceae. Terrestrial plants, herbaceous, with bulbs. Leaves borne on the lower parts of the stems. Flowers in heads, umbels or solitary, with spathes, actinomorphic. P(6), A6, G(3), totally inferior; ovules *n*, axile. Seeds black. *Southwest and central Asia.*

A single genus with 4 species.

Genera inlcuded: ***Ixiolirion.***

233 (297) Tecophilaeaceae. Herbs with corms or tubers. Leaves linear or ovate, borne on the stem. Flowers usually solitary, bisexual, actinomorphic. P(6), A6 or 3+ 3 staminodes, anthers opening by pores. G(3), half-inferior, ovules numerous, axile. Capsule. *Subtropical and tropical America, South Africa.*

There are 6 genera and about 20 species. *Tecophilaea cyanocrocus*, from Chile but now apparently extinct there, is widely grown in Europe for the sake of its blue flowers.

Genera included: ***Odontostomum, Tecophilaea.***

234 (298) Hypoxidaceae. Herbs with rhizomes/corms. Leaves usually basal, alternate, often pleated and hairy; ptyxis conduplicate or plicate. Flowers solitary/racemes/heads, bisexual, actinomorphic. Perianth clearly 3+3, those of the outer series usually hairy outside, A6, G(3), inferior; ovules *n* per cell, axile. Capsule/berry. *Scattered.*

Often included within the Amaryllidaceae (above), there are 5 genera with about 150 species. A few are cultivated as ornamentals.

Genera included: ***Curculigo, Hypoxis, Rhodohypoxis.***

235 (299) Velloziaceae. Shrubs with forked branches with persistent leaf-bases, or woody-based herbs. Leaves alternate, leathery; ptyxis conduplicate. Flowers solitary, terminal, on naked stalks borne in terminal tufts of leaves. P6/(6), petal-like, A6/more in 6 bundles, G(3), inferior; ovules *n*, axile. Fruit a hard capsule, often with spines or glandular. *Tropical Arabia, Madagascar, Africa, South America.*

Two genera (sometimes further divided) and about 200 species. A few are cultivated as glasshouse ornamentals.

Genera included: ***Barbacenia.***

236 (300) Taccaceae. Herbs with scapes. Leaves all basal, broad, long-stalked. Flowers in umbels, each umbel with an involucre of bracts the inner of which are often dangling. P(6), more or less petal-like, A6, G(3), inferior; ovules *n*, parietal. Berry/capsule. *Tropics, China.*

A single genus with about 10 species, a few of which are cultivated.

Genera included: ***Tacca.***

237 (301) Dioscoreaceae. Climbers with swollen rootstocks, sometimes with aerial stem-tubers. Leaves borne on the stems, usually alternate, stalked, often cordate/palmate; ptyxis flat or conduplicate. Flowers in axillary racemes, unisexual, actinomorphic, small. P6/(6), often greenish, A6/3/(6)/(3), G(3) inferior;

ovules 2 per cell, axile. Capsule/berry. *Mainly tropical and warm temperate areas.*

Six genera and over 600 species. Three genera are native to Europe, 2 to North America. Many species are grown as vegetables (yams) in the warmer parts of the world, and a few species are grown as ornamental climbers.

Genera included: ***Borderea, Dioscorea, Rajania, Tamus.***

238 (302) Pontederiaceae. Aquatic herbs. Leaves alternate, with sheathing bases, often stalked. Inflorescence racemose, borne in the axil of a spathe-like sheath. Flowers bisexual, actinomorphic/rarely zygomorphic. P(6), petal-like, A usually 6, G(3), superior; ovules 3–*n*, axile/parietal. Capsule. *Tropics and warm temperate areas.*

Nine genera, 5 of which are native to North America, 4 of these introduced in Europe, and about 30 species of aquatic herbs. Species of about 5 genera are cultivated. *Eichhornia crassipes* has become a noxious weed in many tropical river systems.

Genera included: ***Eichhornia, Eurystemon, Heteranthera, Monochoria, Pontederia.***

239 (303) Iridaceae. Terrestrial herbs with rhizomes/corms/bulbs, rarely woody. Leaves alternate, often equitant and/or pleated; ptyxis usually conduplicate or plicate, rarely supervolute. Flowers solitary/in racemes/panicles, bisexual, actinomorphic/zygomorphic. P6/(6), petal like, sometimes in 2 dissimilar whorls, A3 on the same radii as the outer perianth-lobes, G(3), inferior; styles 3, often further divided; ovules few–*n*, usually axile. Capsule. *Widespread, particularly well developed in southern Africa.*

About 70 genera and 1500 species. Nine genera (1 introduced) are found in Europe and 18 are native to North America. Species of many genera are cultivated for ornament.

Genera included: ***Alophia, Anomatheca, Aristea, Babiana, Belamcanda, Chasmanthe, Crocosmia, Crocus, Cypella, Dierama, Dietes, Diplarrhena, Eleutherine, Ferraria, Freesia, Galaxia, Geissorhiza, Gelasine, Gladiolus, Gynandriris, Herbertia, Hermodactylus, Hesperantha, Iris, Ixia, Lapeirousia, Libertia, Melasphaerula, Moraea, Nemastylis, Neomarica, Nivenia, Orthrosanthus, Pardanthopsis, Rigidella, Romulea, Schizostylis, Sisyrinchium, Solenomelus, Sparaxis, Sphenostigma, Syringodea, Tigridia, Trimezia, Tritonia, Watsonia.***

239a (304) Burmanniaceae. Small herbs, sometimes saprophytic and without chlorophyll. Leaves basal and on the stems. Flowers in cymes, bisexual, actinomorphic. P(3+3)/(3), sometimes 3-winged or -crested to the base, A3, on the same radii as the inner P-lobes, G(3); ovules *n*, axile/parietal; style 1, divided into 3 stigmas. Capsule. *Tropics, extending north to the eastern USA.*

There are 22 genera, with about 130 species; 4 genera are native to North America.

Genera included: ***Apteria, Burmannia, Gymnosiphon, Thismia.***

240 (305) Juncaceae. Herbs, often rhizomatous, rarely woody. Stems terete in section. Leaves usually basal, spirally arranged, sometimes reduced to sheaths; ptyxis supervolute. Flowers in cymes/panicles/corymbs/heads, usually bisexual, actinomorphic. P6, A usually 6, pollen in tetrads, G(3), superior; ovules 3–*n*, axile/parietal. Capsule. *Widespread.*

A family of 9 genera and 350 species. Two genera are native to both Europe and North America; species of both of them are occasionally grown.

Genera included: ***Juncus, Luzula.***

241 (306) Bromeliaceae. Herbs, often epiphytic. Leaves mainly basal, often spiny-margined and/or with elaborate scales covering the whole surface; ptyxis flat or supervolute. Flowers usually in terminal racemes/panicles, with conspicuous bracts, bisexual, usually actinomorphic. K3, C3/(3), A6, anthers usually versatile, G(3), usually inferior; ovules *n*, axile. Berry/capsule. *Mainly tropical America.*

A family of 46 genera and 2100 species of mainly epiphytic herbs. Ten genera are native to North America. About 80 species belonging to 17 genera are cultivated. *Ananas* provides the pineapple of commerce.

Genera included: ***Acanthostachys, Aechmea, Ananas, Billbergia, Bromelia, Canistrum, Catopsis, Cryptanthus, Dyckia, Fascicularia, Fosterella, Guzmania, Hechtia, Hohenbergia, Neoregelia, Nidularium, Pitcairnia, Puya, Quesnelia, Tillandsia, Vriesea.***

242 (307) Commelinaceae. Terrestrial herbs. Leaves mostly borne on the stems, alternate, with closed basal sheaths; ptyxis involute or supervolute. Flowers in panicles/coiled cymes/rarely solitary, bisexual, actinomorphic/zygomorphic. K3, C3/rarely fewer, A3–6, anthers basifixed, filaments often hairy, staminodes 0–3, G(3), superior; ovules few, axile. Capsule. *Tropics and warm temperate areas.*

There are about 35 genera and 600 species. Thirteen genera are native to North America and species of 3 are naturalised in Europe. A few speceis are grown as ornamentals and house-plants.

Genera included: ***Callisia, Cochliostema, Commelina, Cyanotis, Dichorisandra, Geogenanthus, Gibasis, Murdannia, Siderasis, Tinantia, Tradescantia, Tripogandra, Weldenia.***

243 (308) Mayacaceae. Aquatic herbs. Leaves alternate, borne on the stems, slender, apices 2-toothed. Flowers axillary, bisexual, actinomorphic. K3, C3, A3, anthers opening by pores, G(3), superior; ovules several, parietal. Capsule. *Tropical and subtropical America, Southwest Africa.*

A single genus with 4 species, 2 of them native to North America.

Genera included: ***Mayaca.***

244 (309) Xyridaceae. Terrestrial or marsh plants with mostly basal, narrow leaves. Flowers borne in a head, with bracts, bisexual, more or less zygomorphic. K usually 3, segments of differing forms, C(3), A3+0–3 staminodes, G(3) superior; ovules *n*–few, parietal. Capsule. *Mainly tropics.*

A family of 5 genera and 250 species; one genus is native to North America and is occasionally seen in cultivation.

Genera included: ***Xyris.***

245 (310) Eriocaulaceae. Usually marsh plants. Leaves basal or alternate on the stems, sheathing, narrow. Flowers in heads subtended by involucres, unisexual, actinomorphic/zygomorphic. P4–6/(4–5), clearly in 2 whorls but the segments of both whorls similar, all hyaline/membranous, A4–6, G(2–3), superior; ovules 1 per cell, axile. *Mainly tropics.*

Fourteen genera and about 1200 species, mostly from damp places. One genus is native to western Europe and 3 to North America.

Genera included: ***Eriocaulon, Lachnocaulon, Paepalanthus, Syngonanthus.***

246 (311) Gramineae/Poaceae. Herbs/bamboos. Leaves in 2 ranks, alternate, sheathing and with ligules; ptyxis conduplicate or supervolute; stems terete in section, internodes usually hollow. Flowers each compressed between a bract (lemma) and bracteole (palea, rarely absent), the unit forming a floret, these arranged in 2 ranks in spikelets subtended by 2/rarely 1 empty bracts (glumes); spikelets themselves grouped in more complex inflorescences, usually spikes/racemes/panicles. P represented by 2–3 lodicules (often very small), A3/rarely 2/rarely 6 or more, G1-celled, superior, styles 2/rarely 3 or 1; ovule 1, usually lateral. Seed fused to pericarp to form a caryopsis. *Widespread*.

Economically the most important family of flowering plants, with about 600 genera and 9000 species. One hundred and fifty-five genera are native to Europe and 231 to North America. Many are cultivated as ornamentals, and many for their edible grains, e.g. wheat (*Triticum*), oats (*Avena*), barley (*Hordeum*), millet (*Sorghum*), rice (*Oryza*), etc.

Genera included: ***Aegilops, Aegopogon, Aeluropus, ×Agroelymus, ×Agrohordeum, ×Agropogon, Agropyron, Agrostis, Aira, Airopsis, Allolepis, Alopecurus, ×Ammocalamagrostis, Ammochloa, Ammophila, Ampelodesmos, Amphicarpum, Andropogon, Anthaenantiopsis, Anthephora, Anthoxanthum, Antinoria, Apera, Arctagrostis, Arctophila, Aristida, Arrhenatherum, Arthrostylidium, Arundinaria, Arundinella, Arundo, Avena, Axonopus, Bambusa, Beckmannia, Blepharidachne, Blepharoneuron, Bothriochloa, Bouteloua, Brachiaria, Brachyelytrum, Brachypodium, Briza, Bromus, Buchloe, Calamagrostis, Calamovilfa, Catabrosa, Catapodium, Cathestecum, Cenchrus, Chaetopogon, Chasmanthium, Chimonobambusa, Chionochloa, Chloris, Chrysopogon, Chusquea, Cinna, Cleistogenes, Coelorachis, Coix, Coleanthus, Colpodium, Cornucopiae, Cortaderia, Corynephorus, Crithopsis, Crypsis, Cutandia, Cymbopogon, Cynodon, Cynosurus, Dactylis, Dactylotenium, Danthonia, Dasypyrum, Dendrocalamus, Deschampsia, Desmazeria, Diarrhena, Dichanthium, Digitaria, Dissanthelium, Dissochondrus, Dupontia, Echinaria, Echinochloa, Ehrharta, Eleusine, Elionurus, ×Elyhordium, Elymus, ×Elysitanion, Enneapogon, Eragrostis, Eremochloa, Eremopoa, Eremopyrum, Eriochloa, Eriochrysis, Erioneuron, Euclasta, Eustachys, Festuca, ×Festulolium, Garnotia, Gastridium, Gaudinia, Glyceria, Gymnopogon,***

Gynerium, Hackelochloa, Hainardia, Hakonechloa, Helictotrichon, Hemarthria, Heteropogon, Hierochloe, Hilaria, Holcus, Hordelymus, Hordeum, Hymenachne, Hyparrhenia, Hystrix, Ichnanthus, Imperata, Indocalamus, Isachne, Ischaemum, Koeleria, Lagurus, Lamarckia, Lasiacis, Leersia, Leptochloa, Leptocoryphium, Lepturus, Leymus, Lithachne, Lolium, Luziola, Lycurus, Lygeum, Melica, Melinis, Mibora, Microchloa, Micropyrum, Milium, Miscanthus, Molinia, Monanthochloe, Muhlenbergia, Munroa, Narduroides, Nardus, Neostapfia, Neyraudia, Olyra, Oplismenus, Orcuttia, Oreochloa, Oryza, Oryzopsis, Panicum, Pappophorum, Parapholis, Paspalidium, Paspalum, Pennisetum, Periballia, Phacelurus, Phaenosperma, Phalaris, Pharus, Phippsia, Phleum, Pholiurus, Phragmites, Phyllostachys, Pleuropogon, Poa, Polypogon, Psathyrostachys, Pseudosasa, Psilurus, Puccinellia, ×Pucciphippsia, Redfieldia, Reimarochloa, Rostraria, Rottboellia, Saccharum, Sacciolepis, Sasa, Schedonnardus, Schismus, Schizachne, Schizachyrium, Schizostachyum, Sclerochloa, Scleropogon, Scolochloa, Scribneria, Secale, Semiarundinaria, Sesleria, Setaria, Shibataea, Sinarundinaria, Sitanion, ×Sitordeum, Sorghastrum, Sorghum, Spartina, Sphenopholis, Sphenopus, Spodiopogon, Sporobolus, Stenotaphrum, Stipa, Stipagrostis, Swallenia, Taeniatherum, Themeda, Torreyochloa, Trachypogon, Tragus, Tricholaena, Trichoneura, Tridens, Triplachne, Triplasis, Tripogon, Tripsacum, Trisetaria, Trisetum, Triticum, Uniola, Vaseyochloa, Ventenata, Vetiveria, Vulpia, Vulpiella, Wangenheimia, Willkommia, Zea, Zingeria, Zizania, Zizaniopsis, Zoysia.

247 (312) Palmae/Arecaceae. Trees/shrubs/prickly woody scramblers. Leaves large and pleated, becoming pinnately/palmately lobed or divided; ptyxis plicate. Inflorescence a fleshy panicle or spike, often the whole subtended by large, hard, spathe-like bracts. Flowers unisexual/bisexual, actinomorphic. P6/(6), fleshy, A usually 6/rarely more, G usually (3)/rarely 3, 1–3-celled; ovules 3. Berry/drupe, sometimes very large. *Mainly tropics.*

A family of 212 genera and about 2700 species of characteristic habit and appearance (palms). Two genera are native to Europe, 16 to North America. Species of about 30 genera are grown as ornamentals in Europe, particularly in the south.

Genera included: ***Acoelorrhaphe, Acrocomia, Aiphanes, Allagoptera, Archontophoenix, Arenga, Brahea, Butia, Calamus, Chamaedorea, Chamaerops, Clinostigma, Coccothrinax, Cocos, Copernicia, Cryosophila, Cyrtostachys, Elaeis, Gaussia, Hedyscepe, Howea, Jubaea, Licuala, Livistona, Nanorrhops, Phoenix, Prestoae, Pritchardia, Pseudophoenix, Ptychosperma, Rhapidophyllum, Rhapis, Rhopalostylis, Roystonea, Sabal, Serenoa, Thrinax, Trachycarpus, Washingtonia.***

248 (313) Araceae. Herbs/woody climbers, sap often bitter, usually milky; rarely the plants aquatic. Leaves usually stalked and broad, often lobed, veins forming a network; ptyxis supervolute. Flowers minute, stalkless on a spadix which is subtended by and often enclosed in a conspicuous spathe, unisexual/bisexual, actinomorphic. P4–6/(4–6)/rarely 0, A2–8, G1–*n*-celled, superior/naked; ovules 1–*n*, axile/apical/basal/parietal. Fruit usually a berry. *Mainly tropics, less frequent in temperate areas.*

A family of 115 genera and 2000 species. Ten genera are native to Europe, 21 to North America. Species of about 40 genera are cultivated for ornament; several are food plants in the tropics.

Genera included: ***Aglaonema, Alocasia, Ambrosina, Amorphophallus, Anthurium, Anubias, Arisaema, Arisarum, Arum, Biarum, Caladium, Calla, Callopsis, Colocasia, Cryptocoryne, Dieffenbachia, Dracontium, Dracunculus, Epipremnum, Homalomena, Lagenandra, Lysichiton, Monstera, Montrichardia, Nephthytis,***

Orontium, Philodendron, Pinellia, Pistia, Rhaphidophora, Rhektophyllum, Rhodospatha, Sauromatum, Schismatoglottis, Scindapsus, Spathicarpa, Spathiphyllum, Stenospermation, Symplocarpus, Synandrospadix, Syngonium, Xanthosoma, Zantedeschia.

248a (314) Acoraceae. Herbs, aquatic or growing in damp places; all parts with a spicy aroma when crushed, without milky sap. Leaves all basal, narrow, parallel-veined. Inflorescence a spadix borne on a scape, the spathe appearing as a continuation of the scape so that the spadix projects laterally. Flowers bisexual, actinomorphic. P6, A6, G(2–3); style sessile. Berry. *South and east Asia, southern North America.*

A single genus, native to North America, introduced in Europe, with 2 species. One species is commonly cultivated.

Genera included: ***Acorus.***

249 (315) Lemnaceae. Small aquatic herbs, often floating, the plant body not differentiated into stem and leaves. Inflorescence minute, in a pocket on the margin of the plant body and consisting of 1 or more male flowers, each with 1–2 stamens, and a single female flower with a solitary ovary; perianth absent in flowers of both sexes. *Widespread.*

There are 6 genera and about 30 species. Three genera are native to Europe, 4 to North America. A few species are grown in aquaria.

Genera included: ***Lemna, Spirodela, Wolffia, Wolfiella.***

250 (316) Pandanaceae. Dioecious trees/shrubs, often with stilt-roots. Leaves crowded, leathery, keeled, often with spiny margins; ptyxis conduplicate. Flowers in panicles or on spadices, unisexual. P rudimentary/0, A*n*, G1-celled, superior/naked; ovules 1–*n*, basal/parietal. Fruit a syncarp, the woody or fleshy individual fruits fusing. *Old World tropics, Hawaii.*

Three genera and 675 species of striking woody plants. Two genera are native to North America. Only a few species are grown as ornamentals or curiosities in Europe, generally in large glasshouses.

Genera included: ***Freycinetia, Pandanus.***

251 (317) Sparganiaceae. Emergent aquatic herbs. Leaves narrow, alternate. Flowers in unisexual, spherical heads. Perianth of a few scales, A3 or more, G1-celled, superior, more or less stalkless; ovule 1, apical. Fruit drupe-like. *North temperate areas, Australia.*

A single genus with about 12 species; it is native to both Europe and North America. This family is often included in the next.

Genera included: ***Sparganium.***

252 (318) Typhaceae. Marsh herbs. Leaves narrow, alternate. Inflorescence of 2 unisexual, dense, superimposed spikes. Perianth-segments thread-like/scales, A2–5, G1-celled, superior on a hairy stalk; ovule 1, apical. Fruit dry. *Widespread.*

A single genus native to both Europe and North America, with about 15 species, occasionally cultivated for ornament along rivers or streams.

Genera included: ***Typha.***

253 (319) Cyperaceae. Herbs, sometimes large. Stems terete/3-angled in section, usually solid. Leaves spirally arranged, with closed sheaths, ptyxis conduplicate. Flowers subtended by membranous bracts (glumes) spiral/in 2 ranks in spikes/spikelets, without empty glumes at the base, unisexual/bisexual. P scales/bristle-like hairs, A6/rarely 3, with basifixed anthers, G1-celled, superior/naked, sometimes surrounded by a flask-shaped structure (utricle); styles 2–3; ovule 1, basal. Fruit nut-like. *Widespread.*

A large family with over 100 genera and about 4000 species. Twelve genera are native to Europe, 26 to North America. Very few are cultivated for ornament.

Genera included: ***Abilgardia, Blysmus, Bulbostylis, Carex, Cladium, Cyperus, Dulichium, Eleocharis, Eriophorum, Fimbristylis, Fuirena, Gahnia, Kobresia, Lagenocarpus, Lipocarpha, Machaerina, Oreobolus, Remirea, Rhynchospora, Schoenus, Scirpus, Scleria, Uncinia.***

254 (320) Musaceae. Large herbs or apparently small trees, with pseudostems made up of rolled leaf-stalks. Leaves with conspicuous main vein and parallel lateral veins, spirally arranged, as are the bracts. Flowers unisexual, zygomorphic, in spikes, many in the axils of each of the numerous bracts. K3/(3), sometimes joined to the corolla or tubular and split down 1 side, C2-lipped, A5, sometimes with 1 staminode, G(3), 3-celled, inferior; ovules numerous. Fruit a banana. *Old world tropics, introduced into the New World.*

Two genera and about 60 species, including the cultivated banana, *Musa sapientium*. A few species are cultivated for ornament.

Genera included: ***Ensete, Musa.***

254a (321) Lowiaceae. Perennial rhizomatous herbs. Leaves alternate, with conspicuous mid-vein and parallel lateral veins. Flowers in a branched spike, bisexual, zygomorphic. P(3)+3, of which the lowermost is developed as a broad labellum, A5, G(3), inferior; ovules *n*, axile; style 1, stigmas 3, fimbriate. Capsule. *Southeast Asia.*

A single genus with about 6 species, one of them occasionally cultivated.

Genera included: ***Orchidantha.***

254b (322) Strelitziaceae. Herbs/shrubs with leaves/bracts in 2 ranks, leaves with a conspicuous mid vein and parallel lateral veins. Flowers in coiled cymes in the axils of spathes, bisexual, zygomorphic. K3, very unequal, C3/(3), A usually 5+1 petaloid staminode, G(3), inferior; ovules 1–*n* per cell, axile/apical; style 1. Capsule/indehiscent. *Tropics, South Africa.*

There are 5 genera and over 100 species. *Strelitzia reginae*, the bird-of-paradise flower, is widely cultivated and is often used as a cut flower.

Genera included: ***Ravenala, Strelitzia***.

254c (323) Heliconiaceae. Large herbs with rhizomes. Leaves and bracts in 2 ranks, with conspicuous mid vein and parallel lateral veins. Flowers in cymes with large and spectacular bracts, bisexual. K3, C3, no petal forming a labellum, A usually 5+1 staminode, G(3), inferior; ovules 1 per cell, axile/basal. Schizocarp. *American tropics.*

A single genus with about 100 species of large, spectacular, tropical herbs. A few species are cultivated in glasshouses for the sake of their hard, waxy, brightly coloured bracts.

Genera included: ***Heliconia***.

255 (324) Zingiberaceae. Herbs with rhizomes. Leaves in 2 ranks, usually with open sheaths and ligules, aromatic; ptyxis supervolute. Flowers in racemes/heads/cymes, bisexual, zygomorphic. K(3), C3/(3), A1, usually modified into a petal-like labellum, staminodes 3–5, usually petal-like, G(3), inferior, usually surmounted by epigynous glands, style often supported in a groove in the anther; ovules *n*, axile/parietal. Capsule. *Tropics.*

There are about 40 genera and 1000 species. Six genera are native to the southern parts of North America. Species of about 17 genera are grown as ornamentals and species of several more are grown in the tropics as spice plants, e.g. ginger (*Zingiber*), cardamom (*Ellettaria*), turmeric (*Curcuma*), etc.

Genera included: ***Aframomum, Alpinia, Amomum, Boesenbergia, Burbidgea, Cautleya, Curcuma, Elettaria, Etlingera, Globba, Hedychium, Kaempferia, Renealmia, Roscoea, Zingiber***.

255a (325) Costaceae. Perennial herbs with rhizomes. Leaves spirally arranged, with closed sheaths (sometimes opened in later growth), not aromatic. Flowers as for *Zingiberaceae*, but staminodes 3, the laterals absent or reduced to short teeth. *Tropics*.

A family of 5 genera and about 160 species, a few cultivated.

Genera included: ***Costus, Tapeinochilus***.

256 (326) Cannaceae. Herbs with rhizomes. Leaves spirally arranged without ligules; ptyxis supervolute. Inflorescences terminal with flowers in pairs. K3, C(3), A1, petal-like with a half-anther, staminodes several, petal-like, G(3), inferior; style petal-like; ovules *n*, axile. Fruit a warty capsule. *Tropical America*.

A single genus, native to southern North America, with about 50 species, several of which are grown as spectacular, half-hardy, bedding annuals in Europe.

Genera included: ***Canna***.

257 (327) Marantaceae. Herbs with rhizomes. Leaves in 2 ranks, the stalk with a swollen band (pulvinus) at the apex; ptyxis supervolute. Inflorescence a panicle or spike with asymmetric flowers in zygomorphic pairs. K3, C(3), A1 with various petal-like staminodes, G(3), inferior; ovules 1 per cell (occasionally 2 per cell, 1 aborting). Capsule, often fleshy. *Tropics*.

There are 32 genera and about 500 species. Three genera are native to North America and species from 5 genera are grown in glasshouses in Europe for the sake of the frequently coloured or marked leaves. The flowers are individually asymmetric, but generally occur closely associated in pairs, so that the pair is zygomorphic about the vertical axis.

Genera included: ***Calathea, Ctenanthe, Maranta, Stromanthe, Thalia***.

258 (328) Orchidaceae. Terrestrial/epiphytic/saprophytic herbs. Leaves alternate/rarely opposite, often borne on succulent, swollen stems (pseudobulbs); ptyxis conduplicate/supervolute/rarely flat. Flowers solitary or in racemes/panicles, usually bisexual, zygomorphic, usually inverted by a twist in the flower-stalk or ovary. K3/rarely (2–3)/2, C3, all petal-like, the median usually modified into a labellum of varying complexity, A usually 1/rarely 2, united to the style to form a column; pollen in masses (pollinia) of varying shapes and degrees of complexity, G(3), inferior; ovules very numerous, parietal. Capsule, seeds tiny, very numerous. *Widespread.*

Probably the largest family of flowering plants (at least in terms of number of species), with about 800 genera and over 22 000 species. Thirty-five genera are native to Europe and 88 to North America. Many are cultivated, including numerous intergeneric hybrids (some involving up to 8 genera in their formation).

Genera included (*excluding intergeneric hybrids*): ***Acacallis, Aceras, Acineta, Ada, Aerangis, Aerides, Anacamptis, Angraecum, Anguloa, Anoectochilus, Ansellia, Aplectrum, Arethusa, Arpophyllum, Arundina, Ascocentrum, Aspasia, Barkeria, Barlia, Basiphyllaea, Bifrenaria, Bletia, Bletilla, Bollea, Bothriochilus, Brachionidium, Brassavola, Brassia, Broughtonia, Bulbophyllum, Calanthe, Calopogon, Calypso, Campylocentrum, Catasetum, Cattleya, Cephalanthera, Chamorchis, Chiloschista, Chondrorhyncha, Chysis, Cirrhaea, Cleisostoma, Cleistes, Cochleanthes, Cochlioda, Coelia, Coeloglossum, Coelogyne, Comparettia, Comperia, Corallorrhiza, Coryanthes, Corybas, Cryptochilus, Cryptophoranthus, Cycnoches, Cymbidium, Cypripedium, Cyrtopodium, Cyrtorchis, Dactylorhiza, Dendrobium, Dendrochilum, Diaphananthe, Dichaea, Dilomilis, Dimorphorchis, Disa, Domingoa, Doritis, Dracula, Elleanthus, Encyclia, Epidendrum, Epigeneium, Epipactis, Epipogium, Eria, Eriopsis, Erythrodes, Esmeralda, Eulophia, Eulophidium, Eurychone, Eurystyles, Galeandra, Galearis, Gennaria, Gomesa, Gongora, Goodyera, Govenia, Grammangis, Grammatophyllum, Gymnadenia,***

Habenaria, Hammarbya, Harrisella, Helcia, Herminium, Hexalectris, Hexisea, Himantoglossum, Houlletia, Huntleya, Ionopsis, Isochilus, Isotria, Jacquiniella, Koellensteinia, Lacaena, Laelia, Laeliopsis, Leochilus, Lepanthes, Lepanthopsis, Leptotes, Limodorum, Liparis, Listera, Lockhartia, Lueddemannia, Lycaste, Macodes, Macradenia, Malaxis, Masdevallia, Maxillaria, Meiracyllium, Mendoncella, Miltonia, Mormodes, Mormolyca, Nageliella, Neofinetia, Neottia, Neottianthe, Nidema, Nigritella, Octomeria, Odontoglossum, Oecoeoclades, Oncidium, Ophrys, Orchis, Ornithocephalus, Pabstia, Pachyphyllum, Paphinia, Paphiopedilum, Paraphalaenopsis, Peristeria, Pescatoria, Phaius, Phalaenopsis, Pholidota, Phragmipedium, Physosiphon, Piperia, Platanthera, Platyhelys, Pleione, Pleurothallis, Pogonia, Polycycnis, Polyradicion, Polystachya, Ponthieva, Porroglossum, Prescottia, Promenaea, Pseudorchis, Psilochilus, Pteroceras, Pteroglossaspis, Pterostylis, Renanthera, Restrepia, Restrepiella, Rodriguezia, Sarcochilus, Satyrium, Scaphyglottis, Schoenorchis, Schomburgkia, Scuticaria, Sedirea, Serapias, Sigmatostalix, Sobralia, Sophronitella, Sophronitis, Spathoglottis, Spiranthes, Stanhopea, Stelis, Stenoglottis, Steveniella, Tetramicra, Thunia, Tipularia, Traunsteinera, Triaristella, Trichocentrum, Trichopilia, Trigonidium, Triphora, Tropidia, Vanda, Vandopsis, Vanilla, Warrea, Wulfschlagiella, Xylobium, Zuexine, Zygopetalum.

FURTHER IDENTIFICATION AND ANNOTATED BIBLIOGRAPHY

The identification of the family to which a plant belongs is only the first necessary step in its complete identification. To make this book more generally useful, some notes are provided on the most relevant literature that can be used for the purpose. Three broad situations can be defined. These require somewhat different approaches, and are dealt with separately below. Books and papers are referred to by numbers, and are listed numerically (and alphabetically by author) in the bibliography at the end of the chapter.

The family cannot be satisfactorily identified by using the present key. This key does not include all the currently recognised flowering plant families. All exclusively tropical and southern hemisphere families have been excluded unless they contain plants widely cultivated in the northern hemisphere; when this is the case, only those plants cultivated are covered by the key, and other members of these families may well not key out accurately. If the family cannot be identified here, several other works may be used. Some of these have keys (7, 15, 17, 20, 28, 30, 40), others are descriptive, often with illustrations (4, 12, 18, 29, 32, 35, 43). A computer-based interactive key to the families is published on CD-ROM by CSIRO in Australia (42), and computerised versions of 17 are available on the world-wide web.

It is important to re-emphasise here that the circumscription of a particular family can vary from book to book. Care must be taken, therefore, to see that the family arrived at (in this or any other book) corresponds with the family of the same name in yet another work. This can be extremely difficult; it may, however, be done by checking indexes, descriptions, synonyms (where these are given) and comments in the various works against one another. This *caveat* applies to all the works mentioned in this chapter.

The specimen has been identified to its family and its wild geographical origin is known. When the geographical origin of a specimen is known, further identification can be attempted by using a Flora of the region or country in question, if one exists. Floras are too numerous to list here, but details can be found in 5 and 14. If no relevant Flora exists or is available, then the specimen must be treated as though its geographical origin were not known, as below.

The specimen has been identified to family but its wild geographical origin is not known. Under these circumstances, the first step is to find out whether a world-wide monographic study of the family exists. The most notable series of such monographs is that edited by Engler and his successors (11), but this is by no means complete; a newer series of such monographs, recently begun, is available under the general title *Flora of the World* (38). Other monographic studies are published from time to time in various books and botanical journals. Most botanical libraries maintain lists of such publications; many Floras (e.g. 41) contain such lists; and the *Kew Record* (25) includes short abstracts of almost all current relevant publications (since 1971).

Attempts to identify the genus to which a plant belongs can be made by using various works such as 3, 12, 20, 21, 28 and 30, which all attempt to be comprehensive, even though some of them are incomplete. References 6, 16 and 31 list the names of all current genera, with an indication of the families to which they belong (which can be different from book to book, see above).

If the specimen is from a garden plant then it may be possible to identify it using a garden Flora: several of these exist (2, 8, 10, 26, 27, 34, 36, 41).

It is helpful to confirm an identification by comparison with a good illustration. Particularly good illustrations of floral dissections and other organs are provided in 3; and lists of botanical illustrations are included in 22 and 39. In recent years, many popular illustrated works on both wild and garden plants have been produced which, though selective and botanically simplified, can be helpful; volume 3 of 41 contains a bibliography in which many such works are listed.

There are several other books that, though not in themselves usable for identification, contain much useful information. Such works include botanical glossaries, dictionaries, etc. (1, 9, 13, 19, 23, 31, 33).

Finally, the value of comparing the specimen with named herbarium material cannot be overemphasised. This is the most stringent test of the accuracy of an identification, although it should be mentioned that the naming of herbarium material can be wrong or out-of-date. Use of the herbarium is also helpful when the specimen to be named is too incomplete for identification by means of a key.

Bibliographic references

1 Airy Shaw, H. K. (ed.), *J. C. Willis, A Dictionary of the Flowering Plants and Ferns*, 8th edn, Cambridge University Press, 1973.

A valuable source for plant names and, especially, indication of which genera belong to which family; now generally superseded by Mabberley (1997), see below.

2 Bailey, L. H., A *Manual of Cultivated Plants*, 2nd edn, Macmillan, New York, 1949.

A detailed account, with keys, descriptions and illustrations, of the 5000 or so species most commonly cultivated in North American gardens.

3 Baillon, H., *Histoire des Plantes*, 13 volumes (incomplete), L. Guerin, Paris, 1866–95, in French; volumes 1–8 translated into English by Hartog, M., as *Natural History of Plants* 1871–88.

A valuable source of very accurate illustrations (by A. Fauche) of floral and other structures; the text is out-of-date, but still contains observations of interest.

4 Bentham, G. & Hooker, J. D., *Genera Plantarum*, 3 volumes, L. Reeve, London, 1862–83. In Latin.

Now very out-of-date, but still useful for its synopses of genera in each family and its very fine descriptions.

5 Blake, S. F. A. & Atwood, A. C., *Geographical Guide to Floras of the World*. Part 1, Africa, Australasia, N & S America and islands of the Atlantic, Pacific and Indian Oceans, 1942; part 2, western Europe, 1961, both USDA, Washington.
Lists Floras of the areas mentioned up to the dates specified.

6 Brummitt, R. K., *Vascular Plants: Families and Genera*, Royal Botanic Gardens, Kew, 1992.
An extremely useful listing of the families and the genera they contain, as recognised at the Royal Botanic Gardens, Kew; also available at www.rbgkew.org.uk.

7 Cronquist, A., *An Integrated System of Classification of Flowering Plants*, Columbia University Press, New York, 1981.
With excellent descriptions and illustrations of the various families recognised. The keys are synoptic, that is, they do not make any allowance for the numerous exceptional cases.

8 Cullen, J. (ed.), *Handbook of North European Garden Plants*, Cambridge University Press, 2001.
A detailed account (with keys and illustrations) of the families and genera cultivated in gardens in northern Europe.

9 Davidov, N. N., *Botanicheskii Slovar'*, Fizmatgiz, Moscow, 1960.
A multilingual botanical dictionary (Russian, English, German, French, Latin).

10 Encke, F. (ed.), *Parey's Blumengärtnerei*, 2nd edn, Paul Parey, Berlin, 1958. In German.
A taxonomic account, with keys to families and genera, descriptions and illustrations of plants widely cultivated in Germany.

11 Engler, A. *et al.* (eds), *Das Pflanzenreich*, 107 volumes, Akademie Verlag, Berlin, 1900–53. In Latin and German.
A series of family monographs, with keys, descriptions and illustrations; incomplete.

12 Engler, A. & Prantl, K., *Die Natürlichen Pflanzenfamilien*, several volumes, 1887–99; 2nd edn, several volumes, incomplete, Akademie Verlag, Berlin, 1924 onwards.
With keys, descriptions and illustrations.

13 Featherley, H. I., *Taxonomic Terminology of the Higher Plants*, 1959, facsimile edn, Hafner & Co, New York, 1965.

A standard taxonomic glossary.

14 Frodin, D. G., *Guide to Standard Floras of the World*, Cambridge University Press, 1984; 2nd edn, 2001.

The most up-to-date listing of Floras.

15 Geesinck, R., Leeuwenburg, A. J. M., Ridsdale, C. E. & Veldkamp, J. F., *Thonner's Analytical Key to the Families of Flowering Plants*, Leiden University Press, The Hague, Boston, London, 1981.

A very full and complete key, covering all flowering plant families. It uses a different taxonomic system and a very different terminology from that used in this book; the 'Introduction and Notes' should be read carefully before attempting use of the key.

16 Greuter, W. *et al.*, *Names in Current Use for Extant Plant Genera*, W. Koelz, Königstein, 1993. In English.

Lists genus names with indication of the family to which they belong.

17 Hansen, B. & Rahn, K., Determination of Angiosperm families by means of a punched-card system, *Dansk Botanik Arkiv* **26**(1), 1969.

A key to families using easily sorted punched cards; the introduction should be carefully read before use. Computerised versions of this key are available via various sources on the world-wide web.

18 Heywood, V. H. (ed.), *Flowering Plants of the World*, Oxford University Press, 1978 and subsequent reprints. In English.

With descriptions, distribution maps and beautiful illustrations for the various families, but no keys. A new edition is in preparation but is not yet published.

19 Hickey, M. & King, C., *The Cambridge Illustrated Glossary of Botanical Terms*, Cambridge University Press, 2000.

A well-illustrated glossary.

20 Hutchinson, J., *The Families of Flowering Plants*, 2 volumes, edn 2, Oxford University Press, 1959. In English.

With keys to families (and sometimes to the genera within them), descriptions and some illustrations; follows an idiosyncratic taxonomic system.

21 Hutchinson, J., *The Genera of Flowering Plants*, vol. 1, 1964, vol. 2, Oxford University Press, 1967. In English.

An attempt to produce an up-to-date *Genera Plantarum* (see Bentham & Hooker, above), but only 2 volumes were completed; with keys and descriptions for the genera covered.

22 Isaacson, R. T., *Flowering Plant Index of Illustrations and Information*, 2 volumes, G. K. Hall, Boston, MA, 1969.

References to recent articles and illustrations relevant to garden plants.

23 Jackson, B. D., *A Glossary of Botanic Terms*, 4th edn, Duckworth, London, reprinted 1953.

A standard botanical glossary.

24 Judd, W. S., Campbell, C. S., Kellogg, E. A. & Stevens, P. F., *Plant Systematics: a Phylogenetic Approach*, Sinauer, Sunderland, MA, 1998.

A textbook of modern plant taxonomy which provides an explanation for the increase in plant families; family descriptions are included.

25 *Kew Record of Taxonomic Literature*, 1971 and continuing.

Contains abstracts of books and articles of taxonomic interest arranged by families, genera, localities, etc. A consolidated version, which can be interrogated live, is available at www.rbgkew.org.uk.

26 Kirk, W. J. C., *A British Garden Flora*, E. Arnold, London, 1927. In English.

With keys, descriptions and some illustrations to families and genera of plants cultivated in British gardens.

27 Krüssmann, G., *Handbuch der Laubgehölze*, Paul Parey, Berlin and Hamburg, 1962. In German, translated into English by Epps,

M., as *Manual of Cultivated Broad-Leaved Trees and Shrubs*, Timber Press, Portland, OR, 1986.

A very full account of woody plants cultivated in Europe, with some keys, illustrations and descriptions.

28 Kubitzki, K. (ed.), *The Families and Genera of Vascular Plants*, vol. 2 (1993), vols 3 & 4 (1998), Springer-Verlag, Berlin. In English.

A new and continuing attempt to produce a modern *Genera Plantarum* (see Bentham & Hooker, above). Volume 1, published in 1990, covers pteridophytes and gymnosperms.

29 Lawrence, G. H. M., *Taxonomy of Vascular Plants*, Macmillan, New York, 1951. In English.

With descriptions and illustrations of the families and much other information; without keys.

30 Lemée, A., *Dictionnaire Descriptif et Synonymique des Genres de Plantes Phanérogames*, 8 vols, Editions Paul Chevalier, Paris, 1925–43. In French.

With detailed descriptions of families and genera; keys in volumes 8a and 8b.

31 Mabberley, D. J., *The Plant-Book*, Cambridge University Press, 1987; 2nd edn, 1997.

A dictionary of information about plants, with brief descriptions, notes and indication for each genus of the family to which it belongs.

32 Melchior, H. (ed.) *Syllabus der Pflanzenfamilien*, edn 12, Borntraeger, Berlin, 1964. In German.

With good descriptions and illustrations of the families, but no keys. Contains much other matter, including lists of important genera in each family and division of families into subunits (subfamilies, tribes, etc.).

33 Nijdam, J., *Woordenlijst voor de Tuinbouw in Zeven Talen*, 1952.

A polyglot horticultural/ botanical dictionary (Dutch, English, French, German, Danish, Swedish and Spanish).

34 Rehder, A., *Manual of Cultivated Trees and Shrubs*, 2nd edn, Macmillan, New York, 1949.

The classic account, with keys and descriptions, of woody plants cultivated in North America.

35 Rendle, A. B., *Classification of Flowering Plants*, vol. 1, 1930; vol. 2, 1938, Cambridge University Press. In English.

With descriptions and illustrations of the families and numerous informative notes.

36 *The New Royal Horticultural Society Dictionary of Gardening*. 4 volumes, Macmillan, London, 1992. In English.

A dictionary treatment of plants in cultivation; some illustrations, very brief descriptions, occasional keys.

37 Schneider, C. K., *Illustriertes Handbuch der Laubholzkunde*, Gustav Fischer, Jena, 1904–12. In German.

A well-illustrated, very detailed account, with keys, of woody plants cultivated in Europe.

38 Species Plantarum Project, *Flora of the World*, Australian Biological Research Study, Canberra, 1999 and continuing. In English.

A very ambitious project, which is attempting to describe all the plants in the world on a comparative basis. So far, an Introduction (1999) and 10 volumes have been published: 1, Harris, D. J., *Irvingiaceae*, 1999; 2 & 3 deal with Gymnosperm families; 4, Saunders, R. M. K., *Schisandraceae*, 2001; 5, Munro, S. L., Kirschner, J. & Linder, H. P., *Prioniaceae*, 2001; 6, 7 & 8, Kirschner, J., *Juncaceae*, 2002; 9 & 10, Prance, G. T. & Sothers, C. A., *Chrysobalanaceae*, 2003.

39 Stapf. O. (ed.), *Index Londinensis*, 4 volumes and supplement, Oxford University Press, 1921– 41.

A very complete listing of published plant illustrations up to 1941.

40 Takhtajan, A., *Diversity and Classification of Flowering Plants*, Columbia University Press, New York, 1996. In English.

With descriptions of the very large number of families recognised by the author.

41 Walters, S. M. & Cullen, J., *The European Garden Flora*, 6 volumes, Cambridge University Press, 1984–2000.

A very full treatment, with keys, descriptions and some illustrations of all plants widely cultivated in Europe. Volume 3 contains a bibliography, which includes references to many popular, illustrated books on plants.

42 Watson, L. & Dallwitz, M. J., *The Families of Flowering Plants – Interactive Identification and Information Retrieval*. CD-ROM, 1994.

43 Zomlefer, W., *Guide to Flowering Plant Families*, University of N Carolina Press, Chapel Hill and London, 1994.

With descriptions but no keys.

Much useful information, including descriptions and illustrations, can be obtained from sites on the world-wide web by typing a plant family name into a search engine, and studying the results.

GLOSSARY

Only very brief definitions are given here; if more detail is required, reference should be made to the glossaries cited in the bibliography, or to a botanical textbook.

achene: a small, dry, indehiscent 1-seeded fruit; in the strict sense, such a fruit formed from a free carpel.

acicular: needle-like.

actinomorphic: regular, radially symmetric, having 2 or more planes of symmetry. See p. 34.

adnate: joined to an organ of another type (e.g. stamens adnate to corolla); see *connate*.

adventitious (of roots): arising from the stem rather than from other roots.

aestivation: the manner in which the perianth parts are arranged relative to each other in bud. See p. 37.

alternate (of leaves): borne one at each node. See p. 12.

androecium: the male parts (stamens) of a flower, considered collectively.

androgynophore: a common stalk bearing the corolla, stamens and ovary above the sepals.

annual: a plant that completes its life-cycle from seed to seed within a year.

antepetalous (stamens): borne on the same radii as the petals or corolla-lobes and usually of the same number as them.

anther: the pollen-bearing part of the stamen, generally made up of 2 or more elongate sacs. See p. 31.

anthophore: a common stalk bearing the stamens and ovary above the calyx and corolla, as in some Caryophyllaceae. See p. 45.

apical (placentation): see p. 25.

apocarpous: having free carpels. See p. 22.

aril: an appendage borne on the seed, strictly an outgrowth of the funicle. See p. 49.

axil: the upper angle between a leaf and the stem that bears it.

axile (placentation): see p. 25.

axillary: the adjective from axil (see above)

basal (placentation): see p. 30.

berry: a fleshy, indehiscent fruit with the seeds immersed in pulp. See p. 48.

biennial: a plant that completes its whole life-cycle from seed to seed in 2 years.

bifid: shallowly divided into 2.

bilabiate: 2-lipped.

bilaterally symmetric (of a flower or perianth): with a single plane of symmetry. See p. 34.

bipinnate (leaf): a pinnately divided leaf with the leaflets themselves pinnately divided. See p. 14.

biseriate (perianth): in 2 whorls (generally calyx and corolla).

biserrate: regularly toothed, with the teeth themselves more finely toothed.

blade (of a leaf, petal or sepal): the broad, expanded part, borne on a *petiole* or claw.

bole: the trunk of a tree.

bract: a frequently leaf-like organ (often very reduced) bearing a flower, inflorescence or partial inflorescence in its axil.

bracteole: a bract-like organ (often even more reduced) borne on a flower-stalk.

bulb: a complex underground storage organ. See p. 10.

caducous: falling off early.

calyptrate (of a perianth, calyx or corolla) shed as a unit, often in the shape of a cap or candle-snuffer.

calyx: the outer whorl of the perianth, consisting of the sepals. See p. 18.

calyx-lobes: the free parts (equivalent to sepals) of a calyx which has a tube or cup at the base.

capitate: head-like.

capitulum: an inflorescence which is a head (many flowers sessile on a receptacle).

capsule: a dehiscent, usually dry fruit formed from an ovary of united carpels.

carpel: the organ containing the ovules; when several are united, they may be much modified and difficult to distinguish. See p. 22.

caruncle: an outgrowth near the point of attachment (hilum) of a seed. See *elaiosome*.

caryopsis: an achene with the seed united to the fruit wall. See p. 47.

catkin: a unisexual inflorescence of small flowers without petals, with overlapping bracts and often deciduous as a whole. See p. 21.

caudex: the intermediate zone between stem and root. See p. 10.

cauliflory: the bearing of flowers directly on the woody shoots.

cells: The chambers in an ovary of united carpels; also known as loculi.

circinate: see p. 17.

cladode: a lateral, usually flattened, often leaf-like stem-structure borne in the axil of a reduced leaf.

climber: a plant that uses other plants for support.

cluster: an indeterminate inflorescence containing several flowers. See p. 21.

collateral (ovules): borne side-by-side.

compound (leaf): divided into distinct and separate leaflets.

compound (fruit): made up of the products of more than one ovary.

conduplicate: see p. 17.

connate: united to other organs of the same type (e.g. petals connate).

connective: the part of the stamens which joins the anther-cells.

contorted: see p. 37.

cordate (of a whole leaf, or its base): heart-shaped.

coriaceous: leathery and persistent.

corm: an underground stem, very reduced in size and usually vertical.

corolla: the inner whorl of the perianth, made up of petals or corolla-lobes.

corolla-lobes: the free parts (equivalent to the petals) of a corolla that has a tube or cup at the base.

corona: an outgrowth, usually petal-like, of the corolla, stamens or staminodes.

corymb: a flat-topped raceme.

cotyledon: the first seedling leaf (or leaves).

crenate: toothed with blunt or rounded teeth.

cupule: a cup formed from free or united bracts, often containing an ovary or fruit.

cyme: a determinate or centrifugal inflorescence (each axis terminated by a flower).

cypsela: a small, indehiscent, dry 1-seeded fruit formed from an inferior ovary, often loosely termed an achene (see p. 47).

cystolith: a mineral concretion which can be felt in the leaves of some plants.

deciduous (leaves): falling once a year; also used of stipules, catkins, etc.

declinate (stamens, styles): arched downwards and then upwards towards the apex.

decussate (leaves): the opposite leaves of 1 pair at right angles to those pairs above and beneath it.

dehiscence: the mode of opening of an organ, usually an anther or fruit.

dentate: toothed.

diffuse parietal (placentation): see p. 30.

dioecious: with male and female flowers on separate plants.

disc: a fleshy, nectar-secreting organ frequently developed between the stamens and ovary (sometimes also extending outside the stamens).

distichous (leaves): borne alternately on opposite sides of the shoot.

divided (leaves): cut into distinct leaflets.

dorsifixed: attached to its stalk or supporting organ by its back, usually near the middle.

drupe: a fleshy or leathery, 1–few-seeded fruit with a hard inner wall. See p. 48.

drupelet: a small drupe.

elaiosome: an oily appendage borne on a seed, generally near the point of attachment (hilum). See *caruncle*.

endocarp: the inner part of the fruit wall, often hard and stony. See p. 46.

endosperm: food-storage material found in many seeds, formed after fertilisation and incorporating genetic material from the male parent.

entire (leaves): simple and with unlobed and untoothed margins.

epigynous: see p. 46.

epiphyte: a plant that grows physically on another plant, but is otherwise free-living.

equitant (leaves): folded sharply inwards from the midrib, the outermost leaf enclosing the next at the base, etc.

evergreen (leaves): persisting for more than one growing season.

exfoliating (bark): scaling off in large flakes.

exocarp: the outer part of the fruit wall, often forming a rind. See p. 46.

exstipulate: without stipules.

extrorse (anthers): opening towards the outside of the flower.

false fruit: a fruit that includes tissues developed from organs other than the ovary.

false septum: a secondary cross-wall (septum) in an ovary, formed after fertilisation.

false whorl (of leaves): an apparent whorl of leaves produced by extreme shortening of the internodes between the individual leaves (e.g. *Rhododendron*).

fascicle: an indeterminate inflorescence containing more than one flower. See p. 21.

filament: the stalk of the stamen, bearing the anther.

-foliolate: divided into the specified number of leaflets (e.g. 5-foliolate).

follicle: a several-seeded fruit or partial fruit formed from a single carpel, dehiscing along the inner suture.

free-central (placentation): see p. 30.

fruit: the structure containing all the seeds produced by a single flower. See p. 46.

funicle: the stalk of an ovule.

gamopetalous: with the corolla-lobes united at the base.

gamosepalous: with the calyx-lobes united at the base.

gland: a secretory organ.

glume: see p. 249.

gynaecium: the ovary, the female sex organs of a single flower collectively.

gynoecium: alternative (and more frequent) spelling of *gynaecium*.

gynophore: the stalk of a stalked ovary.

half-inferior (ovary): with the lower part of the ovary below the insertion of the perianth and stamens, the upper part above it.

half-parasites: plants which have green leaves but which are also parasitic on other plants.

halophytic: growing in saline soils.

hapaxanthic: see p. 8.

head: see *capitulum*.

herbaceous (of organs): with the texture and colour of leaves.

herbaceous perennial: a plant dying back to soil level or almost so at the beginning of each unfavourable season.

hyaline: translucent and shining.

hypogynous: see p. 46.

imbricate (petals or sepals) overlapping. See p. 37.

imparipinnate: pinnate without a terminal leaflet.

indehiscent (fruit): without any clear opening mechanism.

indumentum: a covering of hairs or scales.

inferior ovary: see p. 40.

inflorescence: the arrangement of flowers on a branch. See p. 19.

internode: that part of a stem between one leaf-base and the next above or below.

intrusive parietal (placentation): see p. 29.

involucel: a cup formed from united bracteoles found below each flower in the heads of species of Dipsacaceae.

involucre: a whorl of bracts beneath an inflorescence.

involute: see p. 17.

labellum: a modified petal or staminode, differently shaped, coloured or sized from the normal petals.

laciniate: deeply slashed into narrow segments.

lamina: the broad part of a leaf or petal; see *blade*.

leaflet: an individual segment of a compound leaf.

legume: a dry, dehiscent fruit formed from a single carpel, dehiscing along both sutures. See p. 47.

lemma: see p. 249.

lepidote: bearing peltate scales; or such scales themselves.

ligule: a tongue-like outgrowth from a petal or at the junction of leaf-sheath and blade.

locules or *loculi*: the cells in a carpel, ovary or anther.

lomentum: an indehiscent several-seeded fruit which fragments transversely between the seeds, forming1-seeded segments.

longitudinal dehiscence (of anthers): opening along the length of the anther.

marginal (placentation): see p. 25.

medifixed (hairs): attached by the middle.

mericarp: a 1-seeded portion of a fruit formed from an ovary of united carpels which split apart at maturity.

-merous: indicating the number of parts (e.g. 3-merous or trimerous).

mesocarp: the central part of the fruit wall, sometimes fleshy.

monocarpic: existing in a vegetative state for several years before flowering.

monoecious: with male and female flowers on the same plant.

multilocular (ovary): with 2 or more cells or loculi.

multiseriate (perianth): a perianth formed from 3 or more whorls of organs.

naked (*ovary*): see p. 40.

nectariferous disc: a nectar-secreting disc within a flower, usually between stamens and ovary.

nectary: a nectar-secreting structure, usually within a flower, occasionally on other parts of the plant.

node: the point on a stem at which a leaf, a pair or a whorl of leaves, is attached.

nut: a hard, indehiscent, 1-seeded fruit.

nutlet: a small nut.

obconical: of the shape of a child's spinning top.

obdiplostemonous (flower): with the stamens twice as many as the petals, those of the outer whorl on the same radii as the petals.

opposite (leaves) leaves borne 2 at each node (generally on opposite sides of the stem).

ovules: the structures within the ovary which become the seeds after fertilisation and ripening.

palea: see p. 249.

palmate (leaves): divided to the base into separate leaflets, all the leaflets arising from the apex of the stalk. See p. 14.

palmatifid (leaves): divided palmately to about halfway from margin to stalk.

palmatisect (leaves): divided palmately to more than halfway from margin to stalk.

panicle: a much-branched inflorescence, strictly a raceme of cymes, but also used for a raceme of racemes.

parallel (veins): veins which are distinct and unbranched from the base of the leaf, running parallel with each other towards the apex.

parasitic: describes a plant which does not photosynthesise (is not green), which obtains all its nutrition from the host plant to which it is attached.

parietal (placentation): see. p. 29.

paripinnate: describes a pinnate leaf without a terminal leaflet.

pedicel: the stalk of a flower.

peduncle: the stalk of an inflorescence.

peltate: disc-shaped, the stalk arising from the middle of the undersurface.

perianth: the outer, sterile whorls of a flower, often but not always differentiated into calyx and corolla.

pericarp: the wall of the fruit, often differentiated into exocarp, mesocarp and endocarp.

perigynous: see pp. 40–6.

perigynous zone: see pp. 40–6.

perisperm: food storage tissue in some seeds, formed entirely from maternal tissue.

petals: the individual segments of the *corolla*.

petiole: the stalk of a leaf.

petiolule: the stalk of a leaflet, (adjective, *petiolulate*).

phloem: tissue within the vascular bundles of the plant which is concerned with the transportation of complex chemicals, only visible with the aid of a compound microscope.

phyllode: a flattened leaf-stalk which takes the place of a leaf.

pinnate (leaves): bearing separate leaflets on each side of a common stalk.

pinnatifid (leaves): divided pinnately to about halfway from margin to midrib.

pinnatisect (leaves): divided pinnately from halfway or more from midrib to margin.

pistillode: a rudimentary, non-functioning ovary.

plicate: see p. 17.

pluricarpellate (ovary): made up of 2 or more carpels.

pollinia: coherent masses of pollen dispersed as units.

polypetalous: with distinct, free petals.

polysepalous: with distinct, free sepals.

pome: a fruit which is made up of an inferior ovary surrounded by fleshy or leathery tissue derived from the receptacle of the flower.

poricidal (anthers): opening by pores.

porose: poricidal.

ptyxis: the manner of packing of the individual leaves inside the vegetative bud.

pyrene: the stone(s) within a *berry* or *drupe*.

raceme: a simple, usually elongate inflorescence with stalked flowers borne individually, the oldest flowers nearest the base. See p. 20.

rachis: the main stalk of an inflorescence or the central axis of a pinnate leaf.

radially symmetric (flower): with several planes of symmetry, *actinomorphic*; see p. 34.

receptacle: the apex of a pedicel, where the floral parts are attached.

replum: a secondary septum formed after fertilisation, in the fruits of most Cruciferae.

reticulate (veins): with veins clearly branching, distinguished into primary, secondary, tertiary, etc.

revolute: see p. 17.

rhizome: underground stem bearing scale-leaves and adventitious roots.

root-tubers: see p. 9.

runner: another name for *stolon*.

saccate (perianth or corolla): with a conspicuous hollow swelling.

sagittate: arrow-head-shaped.

samara: a dry, winged, dehiscent fruit or mericarp, usually 1-seeded.

saprophyte: a plant (with or without chlorophyll) which obtains its food materials largely by absorption of complex organic chemicals from the soil.

scale-leaves: rudimentary leaves borne on a rhizome, or occasionally on stems when true leaves are replaced by *cladodes*.

scape: a leafless inflorescence-stalk arising directly from a rosette (often of basal leaves).

schizocarp: a fruit which splits into separate mericarps.

semi-inferior: half-inferior.

sepal: an individual segment of the *calyx*.

septum (plural *septa*): a cross-wall, generally between adjacent cells of an ovary.

serrate: regularly toothed, saw-like.

sessile: without an obvious stalk.

simple (leaves): not divided into separate leaflets (but possibly toothed or lobed).

solitary (flower): one borne singly at the apex of a stem or scape.

spadix: a fleshy spike of numerous small flowers, generally subtended by a spathe.

spathe: a large bract (or one of a group) which subtends and usually encloses a whole inflorescence in bud.

spike: a raceme-like inflorescence in which each flower is stalkless.

spikelet: a small spike, generally with flowers more or less enclosed between bracts.

spirally arranged (leaves): arranged 1 per node spirally along the shoot.

spur (of perianth or corolla): a nectar-holding or secreting tubular or sac-like projection.

stamen: the male sex organ of the flower, usually consisting of filament, anther and connective.

staminode: a sterile stamen.

stellate (hair): star-shaped.

stigma: the receptive part of the ovary, generally borne at the end of the style, on which the pollen germinates.

stipule: one of a pair of lateral outgrowths arising at the point where a leaf is attached to a stem.

stock: the *caudex*.

stolon: an overground, horizontal stem, generally bearing scale-leaves and rooting at its end.

stones (in fruits): pyrenes.

style: the usually elongate portion at the apex of the ovary, bearing the stigma(s) at its apex.

stylopodium: the swollen joint base of the 2 styles in the ovary of Umbelliferae.

subshrub: plant having persistent aerial shoots near ground-level.

subulate: needle-like.

suffrutescent: having the character of a *subshrub*.

superior ovary: see p. 40.

superposed (ovules): borne one above the other.

supervolute: see p. 17.

syncarp: a multiple fruit, formed from several ovaries.

syncarpous (ovary): with the carpels united.

tendril: a generally touch-sensitive, thread-like organ, coiling around objects touched (rarely with adhesive discs at their ends), providing support for climbing plants.

tepals: the distinct segments of a perianth which is not differentiated into calyx and corolla. See p. 34.

terete: circular in section.

testa: the coat of a seed.

tetrad: a group of 4 pollen-grains shed as a unit.

trifid: shortly divided into 3.

trifoliolate: made up of 3 leaflets.

tripinnate: divided pinnately 3 times.

triquetrous: 3-sided, triangular in section.

truncate: ending abruptly, as though broken or cut off.

tuber: a food-storage organ, generally a modified stem or root, borne underground or above ground.

umbel: a raceme in which the individual flower-stalks all arise from the same point at the top of the inflorescence-stalk.

unicarpellate (ovary): made up of a single carpel.

unilocular (ovary): with a single cavity or cell.

uniseriate (perianth): made up of 1 series of organs.

utricle: a bladdery, indehiscent, 1-seeded fruit, or a sac containing such a fruit.

valvate (sepals or petals) edge-to-edge in bud, not overlapping.

valvular dehiscence (of anthers): opening by flaps or valves.

vascular bundles: tissues concerned with the transport of water and other chemicals, forming a network throughout the plant and particularly conspicuous as forming the major part of the veins in the leaves.

versatile (anthers): attached near the middle and pivoting freely on the filament.

verticillate (inflorescence): the flowers in superposed whorls (verticils), each whorl consisting of 2 opposite, often much modified cymes.

winter-annuals: annuals germinating in the autumn and persisting through the winter as a rosette of leaves, generally flowering early in spring.

xeromorphic: with the habit of plants characteristic of arid regions, e.g. fleshy, or with reduced or fleshy leaves, densely hairy, etc.

xylem: tissue in the vascular bundles concerned with the transport of water through the plant; only visible with the aid of a compound microscope.

zygomorphic: bilaterally symmetric, having only a single plane of symmetry.

INDEX

The authorities for the generic names, which are not given in the main text, are provided here.

Lightning Source UK Ltd.
Milton Keynes UK

176031UK00008B/1/P